ALASKA'S SALMON HATCHERIES

1891- 1959

By

Patricia Roppel

"When all is said and done, and all criticisms on early hatchery procedures and techniques are made, it is true part of these were well founded, but in the main, we made those mistakes due to lack of experience. We were groping in the dark and it simply took time. They didn't send a man to the moon without tremendous research. In the same vein, given the knowledge we have now, scientific and practical, salmon hatcheries, particularly in Alaska where the natural environment hasn't been tampered with, should be highly successful."

Alphonse Kemmerich

Alaska Historical Commission Studies in History No. 20.

The research leading to this publication was financed by the National Marine Fisheries Service. Publication was jointly shared by the Fisheries Rehabilitation, Enhancement, and Development Division of the Alaska Department of Fish and Game; the Auke Bay Fisheries Laboratory of the National Marine Fisheries Service and the Alaska Historical Commission.

Roppel, Patricia
Alaska's salmon hatcheries, 1891-1959 / Patricia Roppel. Portland, Or. : National Marine Fisheries Service, 1982.
x, 299 p. : ill., maps

1. Fish hatcheries--Alaska. 2. Pacific salmon--Alaska. I. United States. National Marine Fisheries Service. II. Title.

SH167.S17 R65 82-600591

Library of Congress #82-600591
ISBN #0-943712-09-2

Printed in U.S.A.

Design/Typesetting/Layout: Drawing Conclusions, Juneau, Alaska
Chinook Salmon Illustration: C.R. Hitz

Contents

Illustrations

Tables

Maps

Acknowledgements

People at various repositories were most helpful as I did research:
Renee Jaussaud and Richard Crawford, at the National Archives in Washington, D.C., were exceedingly helpful in making sure I saw all pertinent papers;
John Kinney, at the Alaska State Archives, made researching at that repository enjoyable and rewarding;
Phyllis DeMuth's staff at the Alaska Historical Library located many sources for me.

I especially appreciate the time spent by those who read all or portions of the manuscript:
Harry Baer — who worked at Afognak hatchery from 1928-1932;
Roger Burrows — who participated in the Little Port Walter experimental hatchery in 1947;
Vincent Boucher — who worked at the Yes Bay and Afognak hatcheries in the late 1920s and early 1930s;
Roger Blackett — who shared research on 1950s Territorial operations in the Kodiak area;
Dr. Robert O. Baker, Robert DeArmond, Richard Harris, J.C. Olsen, Jack Bailey, and Chet Mattson who corrected inconsistencies and provided constructive criticisms.

Others aided me in a variety of ways:
Mark Meyer generously provided logbooks from Quadra and Hetta hatcheries which subsequently were donated to the Alaska Historical Library. He also commented on the portion of the manuscript on the Quadra hatchery, of which he was superintendent;
Mark Morton shared his interest in recording biographies of men who worked in fish culture in federal hatcheries;
Clare Ritter recorded her husband Clifford's remarks on the Quadra hatchery and shared their photographs taken when he worked there.
Frank Roppel, my husband, helped with the maps and many of the sketches and provided much support;
June Zenge typed the manuscript and provided editorial help.
Larry Feeney edited the final draft.

Above all, a special thank you to Alphonse Kemmerich, a warm and generous man who shared his time for an interview, reviewed much of the manuscript, and loaned me his photographs. Kemmerich was superintendent of the Yes Bay hatchery for a number of years and worked at many other federal hatcheries before being promoted to Superintendent-at-large of the Western Region. Under the newly created U.S. Fish and Wildlife Service, he was Regional Supervisor of Fish Culture and then Assistant Regional Director, Region I, and Fisheries Management Coordinator of the Columbia Fisheries Development Plan. After his retirement, he was Executive Director of the Pacific Marine Fisheries Commission. His interpretation of the hatchery activities on the Pacific Coast and in Alaska during those years and his views on today's aquaculture programs provided a perspective on fish culture and hatchery development.

The manuscript was prepared under contract for the Auke Bay Fisheries Laboratory (Auke Bay, Alaska), National Marine Fisheries Service. The contracting officer, Bill Heard, continually encouraged me with his unflagging interest in the subject.

Introduction

The species of Pacific salmon (*Oncorhynchus spp.*) are known popularly by a number of names. The species most propagated in Alaska was *O. nerka*, the sockeye salmon, known as the red salmon in Alaska and blueback on the Columbia River. Other species were *O. tshawytscha*, the chinook salmon, known as the king, spring, tyee or blackmouth; *O. gorbuscha*, the pink, humpback or humpie salmon; *O. kisutch*, the coho or silver salmon; *O. keta*, the chum salmon, also known as the dog or fall salmon. The names used in this book are sockeye, chinook, pink, coho and chum salmon.

Most of the terminology developed to describe the processes and results of fish culture have remained the same over the years. The exceptions are words used to describe young fish. It is the use of the term FRY which causes the most misunderstanding. During the major period of Alaska's hatchery program, 1891-1936, the word FRY referred to what is today's sac-fry or alevin. It is the fish up to the time the yolk sac is absorbed and the fish reach the stage when they swim up or become free swimming and feeding begins. The term ALEVIN for the larval stage of salmon was consciously rejected by the Bureau of Fisheries, as described in Chamberlain, 1907, p. 8. For standardization, the term SAC-FRY has been used to cover this stage of the salmon's development. Anyone reviewing original sources must remember that FRY was synonymous with SAC-FRY.

In this book the following terms are used:
SAC-FRY: the larval stage of salmon
FRY: a fish from the end of the sac-fry period until one year old
FINGERLINGS: fish between one inch and the yearling (or one year old) stage
NUMBER 1 FINGERLINGS: fish from one to two inches in length
NUMBER 2 FINGERLINGS: fish from two to three inches in length

For those unfamiliar with the term BROOD YEAR, a word of explanation will facilitate reading the tables. Brood year refers to the year salmon eggs are taken. Fry released the following spring (in another calendar year) are still called by the calendar year in which the eggs were taken. For example, fry from eggs taken in the fall of 1905 would be termed 1905 brood year despite the fact the fry were released in 1906. To compare number of fry released from the number of eggs taken, use the same brood year for comparison, i.e. 1905 — eggs taken compared to 1905 — fry released.

All tables are compiled from numerous sources — in most cases up to twenty or more references. For ease of reading, sources have generally not been cited in the text. All are listed in the bibliography.

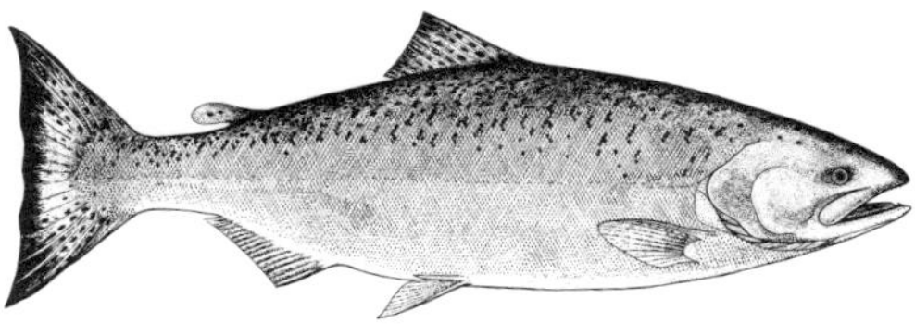

I. Hatcheries to Save Alaska's Fisheries

The sea has been called the great provider, and one of the palatable resources the sea has provided is fish. Chief among these is the salmon. Thousands of streams flowing into the Pacific Ocean from Kotzebue Sound in Alaska to the Monterey Peninsula in California supported runs of Pacific salmon. Since the earliest days of salmon harvests, the danger of exploitation has been apparent.

Throughout the written history of the Alaska salmon fishery, there can be found a constant preoccupation with the threat, and in some instances the fact, of depletion of salmon runs. What could be done to prevent destruction of this natural resource? Limiting the number of salmon taken from the ocean seemed paramount. Over the years since the first Alaskan regulatory legislation on fisheries in 1889, the limitation of the salmon fishery has dealt with one or more of the following variables: limiting the types of gear used to catch salmon; limiting the specific periods of time when it was legal to fish, and the areas where fish could be caught; and limiting the amount of gear allowed to fish. But there is another approach, which is not to limit fishing effort but rather to rebuild or increase the number of fish.

From the early days of Alaska's salmon fishery, there has been a keen interest and enthusiasm for the use of hatcheries as a means of assurance against the adverse effects of commercial fishing. It is thought that the greater survival from egg to fry in the hatcheries over natural survival will produce greater returns of adult fish.

Salmon hatcheries are not a recent development in the fisheries world. Artificial propagation of salmon began in 1763 in Germany, but it took over a hundred years before the first such hatchery was started in the United States. It was in 1864 that the New Hampshire legislature appointed the pioneer fish commission in the United States. Two years later, in 1866, the commission sent Dr. W.W. Fletcher to New Brunswick to procure salmon eggs for the Merrimack River. This was the first attempt ever made in the United States in the direction of salmon breeding. Only two or three hundred fry were actually known to have resulted from the expedition, but it was a beginning which opened a field of operations that has since enlarged beyond the most sanguine expectations.[1]

Five years later, on February 9, 1871, the federal government established by resolution of Congress, the office of the Commission of Fish and Fisheries. Artificial propagation of fishes was not contemplated at that time. The original duties were to scientifically investigate the causes of the decrease in commercial food fish and aquatic animals in coastal and inland waters of the United States and to recommend possible remedies. The following year, at the instigation of the American Fish Culturist Association (forerunner of the American Fisheries Society), an act of Congress authorized the Commission to undertake the propagation and distribution of certain food fishes. The work proved so popular that it was extended annually and soon overshadowed all other work of the Commission.

Fish propagation under the Commission of Fish and Fisheries started on the Pacific Coast with Pacific salmon in 1872 when Livingston Stone took the first chinook salmon eggs on the McCloud River, a tributary of the Sacramento River. From that time, the culture of Pacific salmon spread in California and to Washington and Oregon, primarily to counteract the result of fishing on the Sacramento and Columbia Rivers. As the number of salmon taken on these two major systems decreased, a larger proportion of canned salmon came from Alaska. Large packs were made in 1887 and 1888 on the Karluk River on Kodiak Island in Alaska. The attention of Congress was directed to the inevitable disaster that could overtake the salmon fisheries in Alaska unless prompt means were taken to restrain the destructive methods for the capture of salmon. The first Alaskan salmon fisheries act was passed into law on March 2, 1889, which among other things made it unlawful to use dams, barricades or obstructions in streams to impede the ascent of salmon. It also provided for an investigation into the habits, abundance, and distribution of salmon in Alaska and into the conditions and methods of the fishery. This investigation was conducted that same year, 1889. A group of scientists, under Tarleton H. Bean, an ichthyologist for the Commission of Fish and Fisheries, traveled to Alaska in August. Detailed to the group was Livingston Stone, superintendent of the Commission's California and Oregon hatcheries, who was to inquire into the feasibility of fish culture in Alaska. Since the major salmon fishery was concentrated on Kodiak Island at the Karluk River, the group limited their investigations to that part of Alaska.

Stone examined various sites for suitability of artificial propagation of salmon. The site best suited for a hatchery, Stone felt, was on Afognak Island at Litnik River (now known as Afognak River).[2] Stone was so impressed with Alaska's potential for fish culture, that he recommended to the American Fisheries Society that a National Salmon Park be established. Subsequently he published a paper in *Forest and Stream* and in the annual report of the Commission of Fish and Fisheries.[3] Stone likened salmon to the buffalo and asked for a preserve where the salmon would be unmolested. He recommended Afognak Island and by Presidential Proclamation, December 24, 1892, the entire island was set aside as a Forest and Fish Cultural Reserve. Nothing was done immediately with the reservation except to make it illegal to fish in its immediate waters.

Control of all of Alaska's fisheries, including regulation and hatcheries, has been under various federal agencies. At the time Stone and Bean were examining Alaska's fisheries resources, the U.S. Department of Treasury managed Alaska's fisheries. The Commission of Fish and Fisheries was part of that department. By the late 1890s the Commission shortened its name to the U.S. Fish Commission. Then in 1903 it became the U.S. Commission of Fisheries when Alaskan fisheries came under control of the newly created U.S. Department of Commerce and Labor.

Within the U.S. Commission of Fisheries were several divisions, two of which were active in Alaska: the Bureau of Scientific Inquiry and the Bureau of Fisheries, the latter having the most influence. Within the Bureau of Fisheries was the Alaska Fisheries Service which originally covered only the salmon fishery, but later was extended to all Alaskan fisheries, including the fur seal, in April, 1910. Dr. Barton W. Evermann was one of the chiefs of the Alaska Fisheries Service and had much input into Alaska's hatchery program. In 1913, Labor was split into a separate entity, and the Bureau of Fisheries stayed in the Department of Commerce. The Bureau of Fisheries merged with the Bureau of Scientific Inquiry to form the U.S. Fish and Wildlife Service in June, 1940.

The federal agencies did not build the first Alaskan hatchery. It was the cannery operators who voluntarily pioneered fish culture work in Alaska just 19 years after Stone's first work with Pacific salmon in California. Many companies vied for the masses of salmon which came to the Karluk River each year. It was hoped that a hatchery would keep the numbers of fish at a lucrative level for all. Several cannery operators at Karluk combined forces in 1891, built Alaska's first hatchery, and spawned a few sockeye salmon.

Early fish culture work on the Pacific Coast was generally with chinook salmon, although some coho and steelhead trout were propagated. At the time the first Alaskan hatchery at Karluk began to propagate sockeye salmon, only one British Columbia, Canada, hatchery on the Fraser River had attempted to culture this species. The U.S. Fish Commission did not start propagation of sockeye salmon until 1900, three years after propagation began in the State of Washington.[4]

It is probable that this Karluk hatchery was not the first attempt in Alaska to manipulate the spawning of salmon. There are stories that the Natives of Southeast Alaska moved salmon from one stream to another so more fish or a different species of salmon would return. The Natives of Admiralty Island are said to have spawned salmon and buried the fertilized eggs in the gravel of the same stream in an attempt to prevent fish from digging up the spawn in a crowded environment.[5]

While the details of this Native fish propagation are unrecorded, the first efforts at Karluk proved that sockeye salmon could be propagated, but that propagation procedures were not simple. The hatchery was abandoned but not because of fish culture problems. The cannery men could not agree on fishing rights. An attempt was made to encourage the federal government to take over the hatchery, an appropriate measure in light of its fish culture philosophy.

Alaskan hatcheries began at a time when fish culture was extremely popular as an alternative to restriction of fishing. Why stop fishing when a hatchery would in theory replenish the run? This philosophy was expressed in 1887 by Dr. Brown Goode, the second Commissioner of Fish and Fisheries, and it remained in effect for over 45 years. Dr. Goode believed it was better to expend a small amount of public money to make fish so abundant that they could be caught without restriction and serve as cheap food for the people at large, than to expend a much larger amount to prevent people from catching the few fish that remained.[6]

This philosophy was repeated by a later commissioner, George M. Bowers (1898-1913), who felt that the stocking of waters with food fish was a direct benefit to the public, not only increasing the materials that supported an enormous industry, but providing food for the individual who was willing to use his hook and line.[7] Despite this philosophy and the approval of the Alaskan fisheries agent the Commission of Fish and Fisheries declined to take over the operation of the first Karluk hatchery.

The men in the U.S. Fish Commission who made decisions about Alaska's fisheries and hatcheries obtained information on these subjects from the fisheries agents assigned to Alaska. The Department of the Treasury kept two fisheries agents in Alaska during the fishing season. It was their job to inspect all hatcheries and canneries and to enforce the fisheries regulations throughout the whole of Alaska. Howard M. Kutchin was the most influential fisheries agent during the Treasury's years of involvement with Alaskan hatcheries. His assistant in 1902 and 1903, John J. Coyle, wrote lengthy descriptions of his visits in Southeast as did John N. Cobb, assistant agent from 1904 through 1912. Ward T. Bower was one of the Bureau of Fisheries agents in Alaska for many years and after 1911 he compiled the Bureau's annual reports, *Fishery and Fur Industries of Alaska.*

In addition to the fisheries agents, scientists and investigators also collected information. The Commission of Fish and Fisheries, at the turn of the century, sent Jefferson Moser, a commander in the U.S. Navy, with the Commission's steamer *ALBATROSS* to Alaska to investigate the salmon streams and the salmon industry in Alaska. Since the species most desired was sockeye, the investigation focused on the sockeye producing systems. He was in Alaska in 1897 and again in 1900 and 1901, producing two books of his findings: *The Salmon and Salmon Fisheries in Alaska* and *Alaska Salmon Investigations in 1900 and 1901.*[8] After 1904, Moser joined the Alaska Packers Association (APA) where he became a powerful force because he continued to be considered the authority on Alaskan salmon.

After the Commission of Fish and Fisheries declined to operate the Karluk hatchery, the next attempt to start a hatchery was initiated in 1892 by John C. Callbreath, a private citizen in Southeast Alaska. This was followed by construction of a new hatchery at Karluk (1896) and ones at Klawak Lake (1898) and Tumakof Lake in Redfish Bay (1898), and at Hetta Lake (1899) (Maps 1 & 2). These four were operated by canning companies who wanted to rebuild

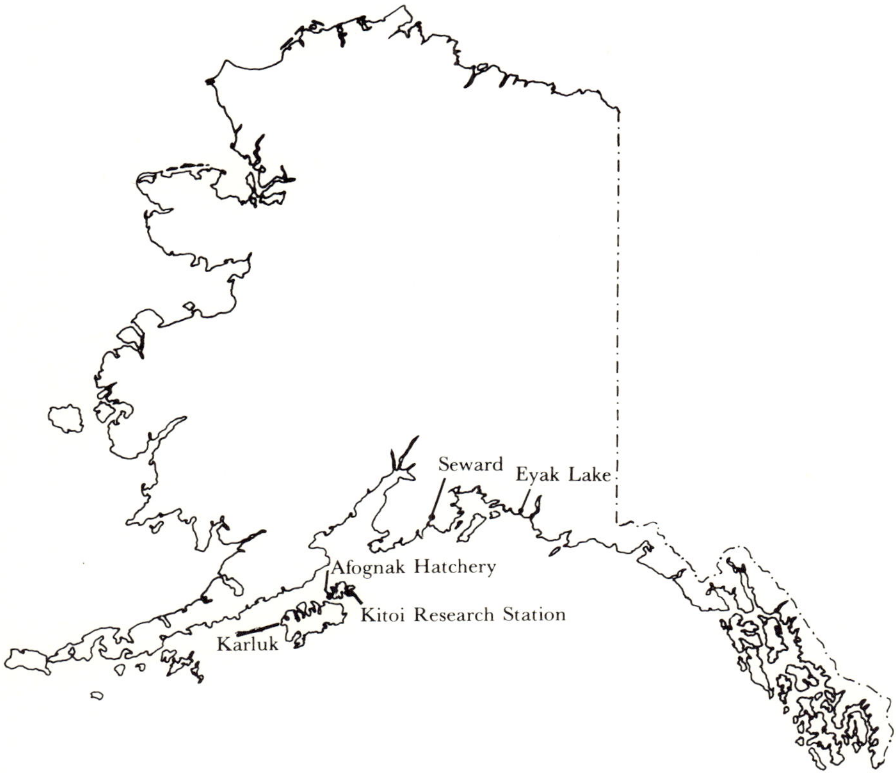

Map 1. Salmon Hatcheries not located in Southeast Alaska

their home streams (streams upon which their canneries were built) to provide a sufficient, steady supply of salmon for their operations.

All but the Redfish Bay hatchery were operating in 1899 when Howard Kutchin, the fisheries agent, wrote in his annual report:

> "The pack has not been maintained, much less increased, because salmon have continued in equal abundance, but because the effort and machinery of the business have been steadily enlarged. Canneries have multiplied, new waters have been explored, and men and gear enormously increased. . . .

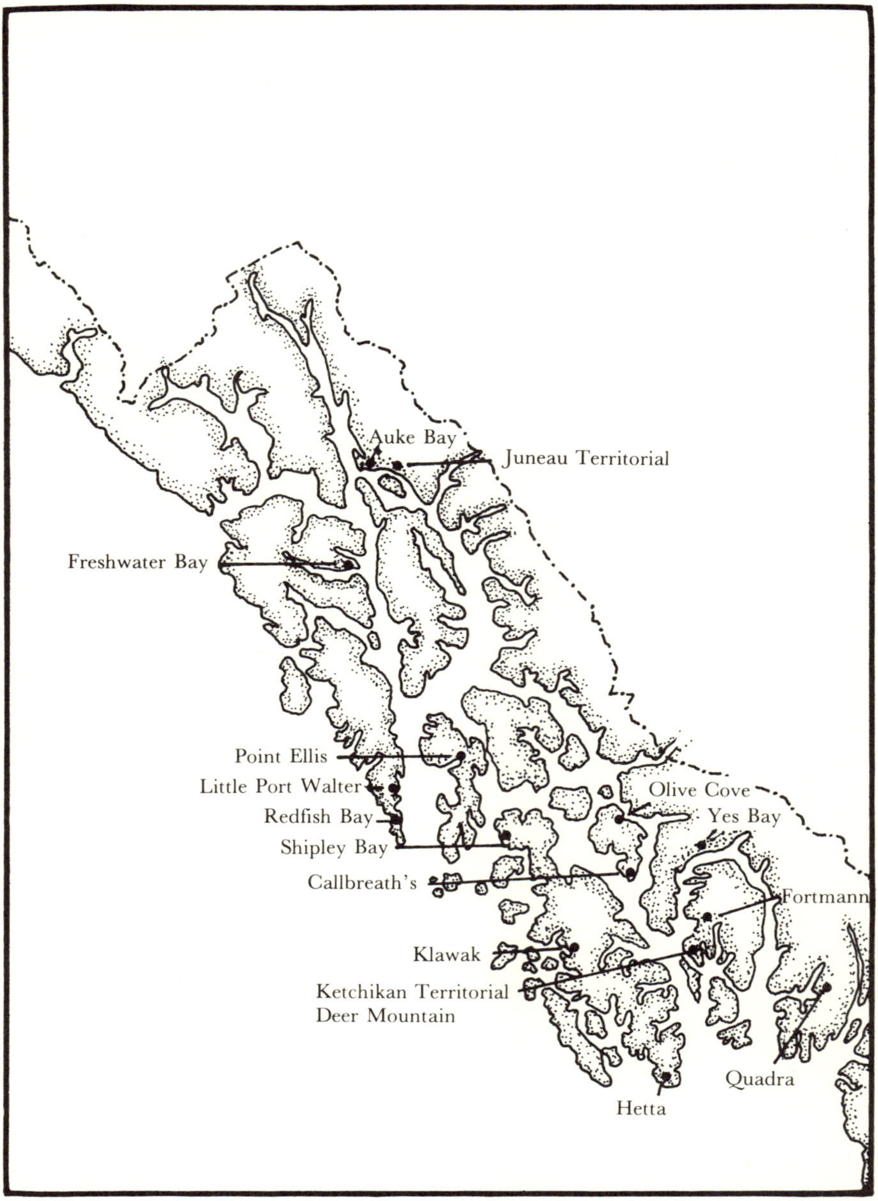

Map 2. Hatchery Locations in Southeast Alaska

> "The uniform conclusion of those who have given investigation and thought to the subject is that the Alaska fisheries are doomed unless swift and thorough measures are put in operation to preserve those which have not yet felt the effect of the destructive practices . . . and to restore those that are rapidly approaching extinction. But one remedy, of course, is available, to-wit: A generous system of propagation, a restriction of the number of canneries, a limitation of the pack at each fishery, and perhaps a curtailment of the fishing season as the differing conditions in the several regions would seem to demand. . . .
>
> "The first branch of this system of protection presents the slightest difficulties and is more readily applicable in a general way than the other elements of the plan. The only question that might present a perplexing phase is as to whether the companies should be required to conduct their own hatcheries or whether the Government should supply the fry from a public establishment, planting the required number at each fishery and collecting the cost from the persons profiting therefrom. My own judgment is that the latter plan would be found the more satisfactory and beneficial."[9]

Kutchin felt that if the government operated hatcheries, Afognak Island could be used for this purpose. He did not think that private voluntary efforts could be relied upon and the three private hatcheries in 1899 were hopelessly inadequate for the necessities of the case.

> "Therefore the Government must take action, either by demanding that each cannery shall yearly replace, in proper ratio, the fish taken by it, or establish hatcheries itself and supply fry at the cost of those who profit from the hatcheries."[10]

Undoubtedly as a result of these suggestions, the 1889 fisheries act was amended on May 2, 1900. This made it mandatory that any person, company, or corporation taking salmon in Alaskan waters must establish a hatchery which would produce sockeye fry in a least four times the number of mature salmon taken each year. A hatchery became mandatory if a packer of salmon wished to stay in business in Alaska. The law did not go into effect until the season of 1901. A few canning companies erected buildings, often on sites where water was of poor quality and quantity, where insufficient numbers of salmon returned to provide eggs for incubation, or where high water destroyed the structures. But a few eggs were hatched and two of the major Southeast Alaska hatcheries were built: the Quadra hatchery at Hugh Smith Lake in Boca de Quadra and the Fortmann hatchery built by the Alaska Packers Association at Heckman Lake. The latter, when enlarged in later years, became the largest hatchery in the world with a capacity of 110 million eggs.

Those salmon packers who complied were in the minority. Most took a wait-and-see attitude. When the assistant fisheries agent, John J. Coyle, arrived in Alaska for his inspection trip of 1902, he was besieged with questions.[11] Was one hatchery sufficient for a company which operated more than one cannery and fished fifty places? Were fry to be put back in all fifty places? Could two or more cannery companies establish a hatchery large enough to produce the required number of fry? The law stated that these hatcheries were subject to rules and regulations. Where were these rules and regulations? Could a company refuse to build a hatchery until the Secretary of the Treasury promulgated these rules and regulations?

Coyle was also asked if the government intended to enforce the law. The question surprised Coyle who said, "I assured him that the government was in earnest and meant to enforce every provision of the law and regulations. . . ."[12] Canneries in business for three or more years were expected to build and operate a hatchery during the season of 1902.

Many of the cannery operators had serious complaints and several argued that the legal requirement to build hatcheries worked a hardship on the small canners.

> "It is, indeed, contended that this class of people would be forced out of business by the strict enforcement of this regulation, and a monopoly created in the interest of the larger concerns that have capital sufficient to conduct hatcheries on a large scale."[13]

Other cannery operators wanted assurance that the government, at some future date, would not erect hatcheries and thereby destroy the value of the property of individuals engaged in the business.

Other arguments were set forth by those who salted salmon. The law was so worded that it specifically required the creation of a hatchery to produce sockeye fry no matter what species of mature salmon were taken. Men who salted pink and chum salmon refused to establish a hatchery for sockeye salmon. Why should a man who made his living on pink salmon put out money and time to enhance sockeye runs which someone else would make money harvesting?

Not only were the cannery operators and salmon salters in disagreement with the law, but so were many of the U.S. Fish Commission's employees. Kutchin still felt that the efforts of private parties in artificial propagation were not satisfactory. He insisted that

> ". . . the efforts of private parties, be they ever so faithful, are not to be relied upon to afford the relief depended on from artificial propagation. It is a business by itself, and an exacting one. The packers are not in the business of raising fish, but that of canning them for market. The manager of a plant has to be a mighty hustler

> to keep up that end of his work, and he simply cannot give the attention necessary to a successful prosecution of the other. The few who are charged with the responsibility now do not profess to know much about it — it is in the hands of an employee who, perhaps, does not know any more than his employer, whatever his professions may be.''[14]

Kutchin was not the only government employee who criticised mandatory hatcheries. Jefferson Moser viewed the act as "highly impractical" and "Localities in the lake regions of the interior (inland) are practically uninhabited and inaccessible and in others the local conditions are not adapted for successful fish culture.''[15] Canneries were located on sites not suitable for hatcheries, especially in northern parts of Alaska where water sources generally froze.

The Commissioner of Fish George M. Bowers dismissed these criticisms and wrote to his superior, the Acting Secretary of the Treasury, in October, 1900:

> "Careful consideration of the matter indicated that it would not be advisable for the Commission to undertake this work at the present time, certainly not until it has been thoroughly tested by the establishment of hatcheries as provided for under Section 7 [1900 act]. . . . It would appear if the present laws are enforced and the canners are required to provide suitable hatcheries, much more will be accomplished than can be done by the government.''[16]

Bowers went on to say that the canning companies could establish and operate hatcheries much less expensively because they had the necessary employees, apparatus to capture fish and available buildings or at least material for such buildings. If the government had to take over artificial propagation of salmon in Alaska, they would have to build hatcheries for all important streams, and they had no transportation or distribution facilities. All of this made hatcheries too expensive for the government.

Mandatory hatcheries for packers of salmon in Alaska remained in effect, but in the strict interpretation of the law, none of these hatcheries came close to releasing four times the number of mature salmon taken. Kutchin had been consulted when the ratio of four to one was proposed, but had been unable to give reliable data with which to set the ratio. He said: ". . . as there is yet so little positive knowledge of the habits of the salmon and no authenticated returns of artificially propagated fish the whole matter is enveloped in uncertainty. Nevertheless, it is better to err on the side of excess. The facility of production once a hatching plant is established, is very great, and the relative cost of additional millions of spawn is triffling.''[17] The ratio of four to one was accepted and became law, but Kutchin was not satisfied. He wrote in the 1900 annual fisheries report: ". . . the ratio fixed in the recent Department order on the subject is now generally believed to be quite below what it should be, and . . . I recommend that it be increased to 50 to 1.''[18] Kutchin was able to convince the Commissioner of Fisheries that an increase in the ratio was necessary and in 1902 the act was amended to read ten times the number of salmon caught each year.

The few private hatcheries which had been built to comply with the law continued to release less fry than required. For example, in 1903 Alaska Packers Association released from its two hatcheries 52,500,000 fish or 113,370,910 less than required as the catch was 16,587,091 adult salmon. Pacific Packing and Navigation Company maintained hatcheries at Hetta and Quadra and released 8,000,000 fish that same year. Since they harvested 8,616,716 fish, the release was less than one-tenth of that required.[19]

An unsuccessful attempt was made at about this time to make mandatory hatcheries more attractive by protecting the return of salmon to hatcheries streams from fishing pressure. The man behind this concept was John C. Callbreath, who, as a private citizen not attached to the fishing industry, ran a hatchery at McHenry Inlet for 13 years. He interested William Sulzer, Congressman from New York, in his concept and Sulzer, in 1902, introduced H.R. 9976, ''to encourage salmon culture in Alaska and for the protection of persons engaged in the production.''

The bill provided that any person or persons who had or would start Alaskan hatcheries would be entitled to the exclusive right of all fish that the hatchery produced in excess of the normal product (today's terminology is escapement) of the stream or barren lake. Callbreath's idea was that streams producing 10,000 or less sockeye salmon or barren lakes could be made to produce large numbers of fish by artificial propagation. The stream was to be protected by prohibiting fishing within one mile of the stream until the normal product or escapement was reached. At that time, commercial fishing would be closed for a further distance of four miles in all directions of tidewater.

Callbreath explained these fishing closures at hearings in 1902 before the Committee on Territories:

> ''. . . the further reserve of 4 miles . . . from which all parties are excluded from taking fish of the kind propagated, will leave those which might be in transit free to pass on to their native stream unhindered, and if they are the product of the protected hatchery, they will go there also unhindered. Now, the object of this is, in many cases there is a little narrow salt-water 'gut' you may call it, running up in the mouth of the hatchery stream and to give a man a mile of that would not be giving him proper protection. Then again, his hatchery might be situated down close to the salt water, where if you give him more than a mile, you would give him a chance to get fish he did not produce, which we do not want.

"Now, if the additional 4 miles is closed to all parties, the hatchery as well as anyone else, if the fish are in transit to another stream, they will go unhindered, and if they belong to him they will come there. . . . Fishing for all kinds of salmon of the different species from those propagated as well as other species of fish, is free and open to all. . . . The propagator is simply given exclusive right to take the fish he produces. All others are free for everyone. . . ."[20]

The rights of Natives were protected by the bill, as they were given free access for food purposes to the hatchery produced fish and had the same rights as others to the natural fish.

The government, through the Secretary of the Treasury, would lease for 20 year periods, subject to renewal, the product of the home stream. The bill did not include provisions for the use of land, only for the natural product of fish. This would then be used for artificial propagation and those fish above this natural product would be used by the hatchery for their purposes, whether it be for sale or for processing in the case of the cannery operated hatchery.

The bill was not supported actively by the canning industry, although they did not oppose it. Fishery Agent Howard Kutchin felt that the bill provided for proper protection of the rights of those who sought to contribute to preserving the salmon runs in Alaska.[21] But the bill was opposed by the chairman of the Committee on Appropriations and several of the members of the Committee on the Territories saw this bill as an establishment of a monopoly. The bill did not pass.

When it became obvious that the mandatory act was unsatisfactory, alternatives were sought. Kutchin remained unshaken in his belief that, although much could be done by private enterprise, the great work could be more scientifically, more cheaply and more thoroughly conducted by the government.[22] This was the viewpoint of many of the cannery operators who felt that the business of propagation of salmon artificially was still in the experimental stages and must be conducted under skilled scientific management and under the most favorable conditions, which could be provided by the government.

Assistant Agent Coyle did not endorse the viewpoint of Kutchin nor of the cannery operators. He wrote in his 1902 annual report:

"I am opposed to paternalism in all its forms, and especially to that form which imposes the taxes upon one class of people for the benefit of another, which is precisely what those persons engaged in the salmon industry ask when they demand that Government hatcheries shall be built and maintained at public expense, for the purpose of preventing the exhaustion of the streams from which they draw their supply. Let the cost of maintaining hatcheries be borne by those who are engaged in taking the fish from the

> streams. If they find it burdensome they will soon enough place it upon the prepared product, thus transferring the burden of increased cost to the consumer of the salmon, where it properly belongs.''[23]

Other alternatives were developed which would not put all of the costs on the government. One was that government establish central hatcheries and sell fry to the cannery operations for distribution. Another was that a special tax be levied and collected on the output of the packers for exclusive maintenance of hatcheries.[24]

By 1903, the controversy was so great that when the Department of Commerce and Labor took over the Alaskan fisheries an Alaskan Salmon Commission was appointed to travel to Alaska on a mission very similar to that undertaken in 1889 by Bean and his group of scientists. The commission, with Chairman David Starr Jordan and vice-chairman Dr. Barton W. Evermann, spent the entire summer and much of the fall of 1903 in Alaska visiting all canneries and every hatchery operation. The fish culturist attached to the group, J.N. Wisner, made detailed notes which he incorporated into a report.[25] Only a brief summary of portions of this was published in the report of the Alaska Salmon Commission.[26]

The conclusion of the Alaska Salmon Commission was that even with its proposed restrictions on methods of fishing, it was questionable whether natural propagation could keep pace with the enormous demands which the canneries made upon the salmon supply. ''The beneficial effects of these restrictions should be supplemented by the establishment of Government hatcheries, if it shall be shown that these can be successfully operated under the very rigorous conditions of the winter climate. It is believed that they can be operated, though at comparatively great cost.''[27] It was suggested that a tax be laid upon the product of the canners and salters and the proceeds applied toward meeting the running expenses of hatcheries. This would be a graduated tax corresponding to the relative market value of the different kinds of salmon packed.[28]

The federal administration, after reviewing recommendations of Kutchin and the Alaska Salmon Commission, was in favor of a plan for government hatcheries, but some of the House of Representative leaders were opposed because of the large appropriations required. The annual estimate was $324,360 for protecting Alaska's salmon, of which $152,000 was for hatcheries and the balance for maintenance, payment of employees and the purchase of two steam launches. When a sundry civil bill passed in 1904, the appropriation did not include the funds for hatcheries.

The following year however, another sundry civil bill, approved March 3, 1905, included funds for the establishment of one or more hatcheries in Alaska for the propagation of salmon at points to be selected by the Secretary of Commerce and Labor. It provided for the purchase of sites, construction of necessary buildings and ponds, purchase and hire of boats, equipment and such temporary help as might be required for construction and operation of hatcheries. Fifty thousand dollars was appropriated, less than a third of the amount requested the previous year.

The site for the first federal hatchery in Alaska was on a lake at Yes Bay in Southeast Alaska. This site took precedence over the Afognak Reserve because of proximity to supply and transportation centers. Construction began in the summer of 1905 and the first eggs were incubated that fall. The second federal hatchery was built in 1907-1908 on Afognak Island at Litnik Lake, near the site selected in 1889 by Livingston Stone and on the reserve established by Presidential Proclamation.

The enforcement of the provision for mandatory hatcheries was found to be impractical so remedial legislation was introduced in 1906. It was drawn by the solicitor of the Department of Commerce and Labor and Mr. Cushman, the House sponsor of the bill, after a number of conferences with the Alaska Salmon Commission, the Alaska fisheries agents, the Secretary of Commerce and Labor, the Commissioner of Fisheries, and a number of representatives of the canning companies doing business in Alaska. The bill as finally passed was in the main satisfactory to the Commissioner of Fisheries and the Bureau of Fisheries and in most respects to the canning companies.

The act of June 26, 1906, superseded the act of 1899 and eliminated mandatory hatcheries. It provided for an inspection for efficiency of private hatcheries in Alaska. This allowed the Commission of Fisheries, through the Bureau of Fisheries, to keep in touch with each hatchery to determine if it was efficient and whether the company was returning to the waters as many sockeye salmon as reported.

The act also introduced a rebate system. There was a tax of four cents a case levied on all salmon canned in Alaska. Hatcheries, by the new act, received a rebate on ten cases, a total of 40¢ for each one thousand sockeye or chinook salmon fry released. It was generally agreed among those drafting the bill that there was an injustice to the private hatchery operators who hatched fish for general benefit without compensation. The plan for rebates was adopted by the government with the view of paying private hatcheries for doing the government's work.[29] However, this did not prove to be sufficient incentive for other canning companies to enter the hatchery program. No new private hatcheries were built because of the 1906 hatchery provision. Two private hatcheries, Hetta and Quadra, were reopened after passage of this law. Whether the rebate system was an incentive to reopen the hatcheries is not known.

The rebate was conditional upon inspection and approval of the hatcheries by the Secretary of Commerce and Labor, through the Bureau of Fisheries. No provision was made by Congress for appointment of a general superintendent or inspector of salmon hatcheries. The superintendent of the Yes Bay hatchery (at that time the only federal hatchery in operation in Alaska) was to inspect each hatchery at least twice a season (once a season at Karluk). He was to determine, among other things, methods of handling and caring for eggs and time, manner and place fish were released to ensure that the maximum number of healthy, vigorous fish were released in a manner to yield best results.

All hatcheries were to be inspected in 1906 before a certification of efficiency would be made, and thus, start the flow of rebates. The Fortmann and Karluk hatcheries of APA were immediately given certification on June 29, 1906, without further inspections because both had been visited and written about so frequently by fisheries agents. The Yes Bay hatchery superintendent was instructed to visit the Klawak hatchery, but the inspection and subsequent certification was not made until February 18, 1907. The Hetta hatchery received certification January 2, 1908, and the Quadra hatchery, January 21, 1910, when operations resumed after an earlier closure due to bankruptcy of the parent company.

The hatcheries were inspected as frequently as the smallness of the Bureau of Fisheries staff in Alaska (three men) and transportation difficulties could permit. The federal hatchery superintendent did not inspect most hatcheries, but rather various Bureau agents made stops when in the vicinity. Some of the best known of these men who made inspections were John N. Cobb in Southeast, E.M. Ball and H.C. Fassett at Karluk and F.M. Chamberlain throughout Alaska. Visits were few and far between. For example, the Karluk hatchery was inspected in July and August 1903 (by Wisner of the Alaska Salmon Commission), June 28 and August 18, 1906, May 7 and September 3-7, 1910 and April 18, 1916.[30] Other hatcheries were visited on a similarly infrequent basis.

Each year the hatchery operators filed sworn statements with the Clerk of the District Court in the First Division for Southeast hatcheries and in the Third Division for the Karluk hatchery. These statements gave the number of sockeye eggs taken and the number of this species released. Duplicates were sent to the Washington, D.C. office of the Bureau of Fisheries. From the number of fry released, the rebate was calculated and a certificate was given which could be applied against any license or taxes on the catch or pack of salmon. No exchange of money was made. The total tax credit given for hatcheries operated during the 30 years was $625,846 (Table 1).

The rebate system was highly criticized. The Grand Jury report made at Valdez in November, 1908 urgently recommend to the Alaska Delegate to Congress that the law be repealed as the taxes which otherwise would be collected rightfully belonged to the Territory for schools and other such purposes.[31]

The Delegate from Alaska, James Wickersham, felt that the rebate system was just another way by which the fishing industry was taking advantage of Alaska. In testimony in 1910 for a proposed amendment to the 1906 act,

Table 1. Hatchery Rebates to Cannery Operated Hatcheries, 1906-1936

Fiscal Year[(a)]	AMOUNT IN DOLLARS					
	Fortmann	Karluk	Quadra	Hetta	Klawak	Total
1906	27,054	11,294	0	0	800	39,148
1907	32,378	14,738	0	0	474	47,590
1908	13,568	17,049[(b)]	0	0	1,110	31,727
1909	9,114	14,842[(b)]	1,210	3,257	1,280	29,703
1910	16,290[(b)]	14,829[(b)]	3,940	3,600	2,120	40,779
1911	16,098[(b)]	16,508[(b)]	4,140	3,421	2,480	42,647
1912	40,134	14,998	4,066	936	1,412	61,546
1913	8,320	16,721	3,250	1,436	1,470	31,197
1914	3,480	12,618	6,821	1,436	1,386	25,741
1915	8,328	11,081	8,120	2,857	1,461	31,847
1916	10,022	9,579	2,836	3,039	1,608	27,084
1917	22,962	0	6,001	1,248	3,128	33,339
1918	2,454	0	5,196	1,834	0	9,484
1919	6,082	0	7,910	0	0	13,992
1920	6,828	0	4,542	0	0	11,370
1921	6,950	0	7,565	0	0	14,515
1922	5,154	0	3,858	0	0	9,012
1923	6,794	0	2,402	0	0	9,196
1924	5,500	0	6,893	0	0	12,398
1925	4,402	0	7,772	0	0	12,174
1926	6,396	0	7,738	0	0	14,132
1927	7,532	0	7,736	0	0	15,276
1928	0	0	7,744	0	0	7,744
1929	0	0	7,736	0	0	7,736
1930	0	0	4,514	0	0	4,514
1931	0	0	8,106	0	0	8,106
1932	0	0	7,761	0	0	7,761
1933	0	0	8,869	0	0	8,869
1934	0	0	8,012	0	0	8,012
1935	0	0	3,944	0	0	3,944
1936	0	0	13,012	0	0	13,012
TOTAL	$265,840	$154,257	$163,956	$23,064	$18,729	$625,846

[(a)] Fiscal year ending June 30.

[(b)] When their hatcheries did not distribute all fry before July 1, Alaska Packers Association carried the remaining fry into the subsequent fiscal year.

Wickersham testified that the Alaska Packers Association in 1907 canned 806,808 cases of salmon and at 4¢ per case, the tax amounted to $32,272.32. The company turned in hatchery certificates in lieu of tax of $32,272 and they paid 32¢ in stamps. "So that they paid no taxes to the Government nor to Alaska, or anybody else, except 32 cents, for the enormous amount of fish that they took out in the season. Of course, they say they turn back the fry into the sea. I want to call attention to that. Nobody knows whether they did or not."[32] Wickersham requested that all future annual Bureau of Fisheries reports should include both the rebates and the taxes so comparisons could be made. Future reports included the rebate but not the taxes paid by each company.[33]

Wickersham's testimony led to the belief that the canning companies were making money operating hatcheries. In hearings in 1912 APA official, Jefferson Moser, was prepared to refute the accusation that his company was using their hatcheries to make money. He explained that the Karluk and Fortmann hatcheries, with the capacity of 160 million eggs, cost from 1896 and 1900 (respectively) to 1911, $475,000 for construction, improvements and maintenance but not depreciation, salaries, transportation and office costs. Rebates for this period, he said, were only $199,800.[34] Even if operating expenses alone were considered, Moser testified, APA's "profit" made on their hatcheries from rebates for the six years prior to 1911 was $43.84, or 0.016 of 1 percent per year.[35]

The original intent of the drafters of the 1906 act was to be certain there was no profit involved and no opportunity for speculation. At that time, the Bureau of Fisheries established the actual cost of producing salmon fry at 40¢ a thousand or $400 per million fry. The average cost per million fry at Fortmann (1905-1911) was $360 and $430 at Karluk.[36] According to Northwestern Fisheries Company official, C.H. Buschmann, the estimated average cost per million fry at his company's Hetta hatchery (1909-1913) was $606, and at the company's Quadra hatchery, $396 per million fry.[37] The cost of operating the government hatcheries, minus costs of construction and maintenance, were similar. In 1910 the cost at the Afognak hatchery was $500 per million.[38] At the Yes Bay hatchery, the cost (1907-1909) averaged $370 per million. None of these figures is representative of the costs of operating hatcheries once fry feeding became standard.[39] The conclusion of those in attendance at the Senate hearings was that private enterprise could operate hatcheries as cheaply or more cheaply than the government. The rebate system simply meant that the government was buying salmon fry.[40]

Yet another criticism of the rebate system was that proof of the number of fry actually liberated was always in question. Since government inspectors were not kept on site and official inspection occurred infrequently, falsification of records could be easy. Alaskan Governor Walter E. Clark, in testimony at the 1912 hearings, felt that this criticism was unjustified:

"A good many times, when reckless statements have been made in

> the country, very often for political purposes, that the present proprietors of private salmon hatcheries in Alaska were robbing the Government and robbing the people, I have had occasion to comment on these statements.
>
> "I have never received affirmative evidence that hatcheries were not being run in good faith and with perfect honesty. There is a penalty for false statements made by the proprietors of the hatcheries or the managers of the hatcheries before the Clerk of the Court. False statements are punishable as perjury is punishable and I have no idea nor belief that there has ever been any dishonesty. . . ."[41]

Bureau of Fisheries officials felt that there was no adequate basis for the belief that large releases of imaginary fry could be reported. Millard Marsh, another of Alaska's fisheries agents, testified in 1910, "The Department knows the capacity of each hatchery and something of the run of salmon each season in the hatchery streams, and this alone would make wholesale frauds almost impossible."[42]

Another criticism of a rebate system which paid for numbers of fry put into the water was that it encouraged poor hatching practices. Condition of the fry and the chance for survival, critics felt, were not important, but only the number of fry released.

A bill, S. 5856, was introduced in 1912 to eliminate private hatcheries and, thus, the rebates, but it did not provide for the government to take over or reimburse the owners of existing hatcheries. The Fortmann and Quadra hatcheries had been built to comply with an act of Congress requiring hatcheries of all persons taking salmon from Alaskan waters. The Karluk hatchery had been rebuilt and enlarged to comply with the act. This meant that over half of the operating private hatcheries were constructed to comply with a federal law. This amendment to the 1906 act was not passed.

Although the distrust and criticism of the rebate system continued, the prevailing attitude was expressed by Moser in the 1912 Senate hearings ". . . a bird in the hand is worth preserving and it were far better for all concerned to encourage private hatcheries by permitting a just rebate at least until such time when the government has actually established and is actually operating a sufficient number of hatcheries to satisfy the drain on the fisheries."[43]

Congressional delegates were reluctant to commit the government to operating what was considered a sufficient number of Alaskan hatcheries to make up for the large numbers of fish caught each year. Moser, in the same Senate hearings, suggested that nine additional government hatcheries were necessary.[44] He gave the following locations for sockeye hatcheries: Chignik,

Cook Inlet, Chilkoot Inlet (head of Lynn Canal), Lake Iliamna and the Wood River in Bristol Bay. Humpback or pink salmon hatcheries should be located, he thought, in Prince William Sound and three at unspecified sites in Southeast. Each of the nine hatcheries would have the capacity for 100 million eggs. Hatcheries were expensive, congressional delegates felt. In 1910 the Bureau of Fisheries estimated the construction cost of each at $40,000. Not only would operating budgets be needed, but the number of government fisheries agents would certainly have to be increased over three men, which also increased the federal appropriation needed. The number of federal hatcheries in Alaska was not increased.

A 1916 bill, HR 9528, to again amend the 1906 fisheries act, included provisions for the Secretary of Commerce to purchase or acquire by condemnation all private hatcheries with certificates to be used for payment of taxes.[45] But the chairman of the House Committee on the Merchant Marine and Fisheries was suspicious and said, "I am a little suspicious that they [the private hatchery operators] are handing us a lemon, however." and he later reiterated his reluctance, "If they are not making any money, the Government will pretty soon go bankrupt taking over all the broken down businesses of the country."[46] Again an amendment to the 1906 act did not pass.

The owners of private hatcheries did not object to government operation of all hatcheries in Alaska. They had been strongly in favor of such a policy since the first hatchery was built in 1891 by private industry. They did, however, expect to be reimbursed for their investment. Many Bureau of Fisheries officials felt that even if the private hatcheries were purchased, they would be insufficient and new hatcheries still would be needed.[47] The difficulty was to persuade Congress to appropriate the money needed for hatcheries.

The cannery operators had a solution which did not prove to be acceptable. In 1912 the APA submitted for consideration an amendment to S. 5856. It read:

> "Provided: that all taxes, licenses, and other fees collected from or through the salmon fisheries of Alaska be used for the construction and maintenance of salmon hatcheries and for construction and maintenance of an experimental station to be used for the development of the fishery product along commercial lines in such a manner as the Secretary of Commerce and Labor may direct."[48]

The Alaskan Delegate, James Wickersham, was very much in agreement with the idea of fisheries taxes being used for hatcheries. At hearings on S. 5856, Wickersham stated:

> "I think that the Government ought to maintain all the hatcheries and that the fisheries ought to be free.... I think Dr. Evermann [chief of the Division of Alaska Fisheries] ought to be given money

> enough and force enough to build hatcheries and maintain hatcheries and keep the fisheries of Alaska up to their highest point of efficiency until the state of Alaska is formed and takes the burden away from him by virtue of the constitution, and that the fisheries then ought to be free.''[49]

Using the taxes from the fishery for hatcheries did not prove to be a popular idea. This amendment to the 1906 act also did not pass.

These proposals to change the provisions in the 1906 act on hatcheries did not affect fish culture operations in Alaska. Few of the fish culturists at their remote locations knew about the controversy as they went about their business of taking salmon spawn, incubating the eggs, and rearing the fry for release. The one exception was the Karluk hatchery.

One of the reasons the Bureau of Fisheries and the Department of Commerce encouraged government take-over of the private operations was because the scientists in the Bureau of Fisheries felt very strongly that the Karluk operation was ineffective. The APA was so politically powerful that the Bureau officials felt they could not close the hatchery unless they had scientific, irrefutable proof that the operation was not effective. This topic will be further developed in the history of the Karluk hatchery. The Karluk hatchery was closed voluntarily in 1917, and no further bills were introduced to eliminate private hatcheries or repeal the rebate system. By 1917 the only private hatcheries in operation were Quadra, Fortmann and Hetta.

At the same time Alaska had become disillusioned with the Organic Act of 1912, which provided territorial status but little home rule. One of the goals was control over the territory's natural resources, such as the fisheries. Governor John F.A. Strong wrote in his annual report in 1917: ''It is submitted that the time has arrived when the salmon and other fisheries of Alaska should be passed to the control of the Territory in all respects save that of scientific direction, which should remain in the hands of the Bureau of Fisheries.''[50]

During the 1917 Alaska legislative session, J.R. Heckman, Ketchikan, introduced S.B. 29 which set aside $80,000 in a fish hatcheries fund, ''to be used for the purpose of building and operating fish hatcheries and the protection and care of natural spawning grounds in the Territory of Alaska.''[51] It set up a board of three fish commissioners appointed by the Governor, with the Governor as ex-officio member. The Commission was to appoint a general hatcheries superintendent. The Governor stated in his annual report that the Territorial Legislature, in making this appropriation, ''clearly demonstrates its willingness to render substantial aid to the conservation and protection of the salmon fisheries.''[52]

The bill did not reach the Governor for his signature until after the legislature adjourned. Then an irregularity in Senate approval of the bill was discovered. The Treasurer and Attorney General of the Territory ruled the funds could not be expended legally.[53] The Territorial program for salmon propagation came to a standstill.

During the 1919 legislature S.B. 10 was passed again appropriating $80,000 "for the purpose of building and operating fish hatcheries; for the protection and care of natural spawning grounds; for the carrying on of fish research work and artificial propagation."[54]

Prior to the creation of the Territorial Fish Commission, the Alaska Fish and Game Club, which was organized in 1917, had been conducting certain experiments in planting salmon and trout eggs in the fertilized and eyed stage in the gravels of streams. A hatchery for the purpose of handling the eggs collected for such experiments was established in Juneau. On the creation of the Territorial Fish Commission in 1919, this experimental hatchery was purchased by the new organization and the available equipment used in the installation of a larger station, also in the city of Juneau. As the active members of the Commission were officers of the Alaska Fish and Game Club, the general policy of the Commission remained unchanged until 1923.

A major reorganization occurred that year when Governor Scott C. Bone, as chairman of the Territorial Fish Commission, appointed J.R. Heckman to the Commission and as vice-chairman. Heckman, superintendent of the APA cannery at Loring until 1918, had been in charge of the Fortmann hatchery operations. The Territorial Fish Commission closed its Juneau hatchery and moved the equipment to a site in Ketchikan. Heckman changed the emphasis from coho and sockeye salmon culture to pink salmon, and from planting eyed eggs to rearing fry. The Commission also operated hatcheries for varying periods of time at Eyak Lake near Cordova and Grouse Lake near Seward. In addition, the Territorial Fish Commission rehabilitated many streams to provide more spawning area. The policy was "Natural obstructions over which salmon have passed since time immemorial should not be molested. Accidental obstructions should be removed whenever practical."[55]

A concentrated predator control program was initiated, including destruction of Dolly Varden and of hair seals on the Copper River flats. It was the intention to destroy predators until the proper equality was reached between the species or the natural balance was regained.

The four hatcheries, only three of which operated at the same time, experimented with coho, chinook and pink salmon propagation as well as the traditionally propagated species, sockeye. Small appropriations curtailed the size of these operations, none of which reached the proportions of Fortmann or the two federal hatcheries.

The success of Alaskan hatcheries was questioned by many. William L. Paul, a powerful Native leader and editor of the *Alaska Fisherman* challenged the

Territorial Fish Commission to prove the value of its hatcheries. He also objected to the composition of the Territorial Fish Commission board as early as 1923.[56] He wanted the legislature to elect the commissioners, instead of selection by the Governor. "As long as the Commission is a political instrument, let us make it a representative one. And if in so doing, it remains a cannery instrument, we will not complain."[57] He was not satisfied with answers received on both topics, so, as a territorial legislator, he introduced in 1927 H.B. 13 and H.B. 14 to repeal the 1917 and 1919 acts. He also introduced H.B. 15 to create a new Territorial Fish Commission and prescribe its duties. The repeals were passed before a new commission was created and heated debate arose when no one could agree on how to set up this new commission. The measure was tabled and the Territorial Fish Commission went out of existence.

R.A. Lomen, a legislator from Nome, felt that the hatchery program should be saved, particularly in Ketchikan where two million chinook salmon fry were being held. He fought long and hard to obtain funds. Finally H.B. 88 authorized the Territorial Board of Road Commissioners to lease and care for all property in possession of the Territorial Fish Commission and to engage A.J. Sprague to carry on experiments in fish culture. Only $8,000 was appropriated.

Karl Theile, secretary of the Territorial Board of Road Commissioners stated in a letter to Sprague, "After carefully considering the Act it is the opinion of the Board that the essential requirement of the law is found in subsection (d)."[58] This meant that Sprague's principal duty was to prepare a detailed report for the Territorial Legislature which included such information as the number of fish of each species taken near every stream in Alaska for each year from 1919 to 1927. Theile added, "No doubt you can arrange your work so that you will have some time to devote to experiments in fish culture."[59] There was very little hatchery work to supervise except at Ketchikan. The Seward hatchery had been destroyed by fire March 6, 1927; the one in Cordova was not operating because the appropriation was exhausted when a new hatchery building greatly exceeded cost estimates.

Sprague's report of 1927-1928 was the last report of the Territorial Fish Commission. The 1929 legislature appropriated no money for continuing reports nor for fish culture. In 1931 H.B. 103 relieved the Territory of any further fish culture responsibilities. It provided for disposition of the fish hatchery buildings and property.

At that time few hatcheries remained in Alaska. The strong interest in fish culture as the savior of depressed salmon runs was no longer there. The private hatchery operators could not see definite results. The runs of salmon to the home stream at the hatcheries did not show great increases of fish. It is impossible to determine if this was caused by poor hatchery practices or because of increased fishing pressure with increased, bigger and more efficient gear, or from inadequate regulations.

As more and more processing plants for canning, freezing and mild curing of salmon were constructed there was more demand for fish. Thus, the pressure to catch more fish increased. An example is the increase in fathoms of purse seine used in Southeast Alaska, where most of the hatcheries were located. In the eight years prior to 1920 the number of fathoms of gear increased 58 percent. During the 1920s the gear stabilized, but increased by 10 percent in the 1930s and another 10 percent in the 1940s. It increased 67 percent in the 1950s. Total increase in fathoms of purse seines from 1912 to 1960 was 257 percent.[60] Not only were purse seines more numerous, they also became more efficient. Seine boats became powered with gasoline engines about 1903, which extended fishing range. Soon a winch was used to bring the purse seine aboard making it possible to haul more sets, thus catching more salmon. The fish traps also increased, particularly in Southeast, and the share of the catch so taken steadily increased. Prior to 1909, traps took less than a quarter of the landings. By the early 1920s this figure had grown to almost 45 percent and in the late 1920s, the peak period of traps, they accounted for over 50 percent of the total catch of all species of salmon. The proportion remained stable throughout the 1930s and declined after that.[61] The gillnet fishery did not have as much influence in the areas of Southeast where hatcheries operated, nor did trolling, which does not harvest significant numbers of sockeye salmon.

The majority of the hatcheries were located in the southern half of Southeast Alaska (from an imaginary line drawn through the archipelago from Petersburg to the Pacific Ocean — what is today's Alaska Department of Fish and Game's Districts 1-8). (Map 3). Catch records have been compiled for this section of Southeast, thus eliminating the big runs of sockeye salmon to Lynn Canal and Taku Inlet where no hatcheries operated.[62] The Karluk and Afognak hatcheries operated in isolated areas and harvest records have been examined in the history of each hatchery later in this book.

Figure 1 shows the commercial catches of sockeye salmon in the southern half of Southeast Alaska where Quadra, Fortmann, Klawak, Hetta, Yes Bay and Callbreath's hatcheries operated. It can be seen that the commercial harvest of sockeye salmon fluctuated wildly. The number of sockeye eggs which could be taken also fluctuated as shown in Figure 2. The releases of sockeye fry, as shown in Figure 3, have little correlation with the commercial harvest of sockeye salmon. It will be noted that after the last of the hatcheries producing sockeye salmon was closed, the sockeye catches in southern Southeast Alaska did not reach the highs which were experienced in the earlier years when the hatcheries operated. Although the loss of hatchery production may have been significant, the decline occurred at a time when seine gear continued to increase. Trap catches, however, decreased and the fishery was more closely regulated because the 1924 White Act required that 50 percent escapement be maintained in each stream.

In the early years there had been little restriction on fishing and early hatchery operators enjoyed little protection of the salmon in their home streams. None

Map 3. Southern Southeast Alaska

of the early laws protected a hatchery's product. What restriction of fishing in Alaska did exist was solely by federal law.

The first major restriction of fishing in Alaska was the act of 1896 which prohibited fishing above tidewater in streams of less than 500 feet in width and prohibited setting gear across tidewater of streams for more than one-third the width or within 100 yards of another net or seine. The 1900 regulations prohibited nets within 100 yards of stream mouths (measured from mean low tide). This prohibition was also included in the 1906 act. All other waters in Alaska were open to fishing except those streams or areas set aside as spawning grounds or fish culture reserves, which were authorized by an 1896 amendment to the 1889 fisheries act.

The Secretary of the Treasury could set aside these fish culture reserves if, in his judgment, the results of fishing operations on that stream indicated that the number of salmon being taken failed to reflect the stream's expected production. But only after a public hearing of all interested parties could fishing be limited or prohibited and the stream become a fish culture reserve.

When the Alaska Salmon Commission examined Alaska's fishing potential in 1903, Wisner suggested 13 streams for propagation of salmon be withdrawn from the public domain, subject to the possessory rights of Natives and persons claiming title through the Russian Government and subject to the rights of Natives to take fish from the waters for individual use. These systems and their catchment basins were the Naha River, all streams flowing into Yes Bay, Klawak Stream, Boca de Quadra stream (Sockeye Creek), Karluk River, Karta River (in Kasaan Bay), White River, Norwiskay River (Miller Creek in Moria Sound), all streams draining into Olga Bay on Kodiak Island, Chignik River, streams at the head of Chilkat and Chilkoot inlets and streams flowing into English Harbor (Lower Cook Inlet).[63]

Nothing was done with these fish cultural withdrawals except for Yes Bay which became a fish cultural reservation by Presidential Proclamation. The APA was in possession of certain of these sites and became technically trespassers on the land. When they began to seek title of those lands around 1909, it was found that in view of the 1906 act, which superseded all previous Alaska fisheries acts, it was questionable if there was authority to make withdrawals for this purpose. No hearings had been conducted as specified. Therefore, it was decided in 1909 that the withdrawals would be revoked and the lands restored to public domain.[64] After 1909 there were no further land withdrawals for fish cultural purposes.

Greater restrictions on fishing were made after this time, and several of the streams upon which hatcheries were operated were closed to fishing within 500 yards of the stream mouths. Other hatcheries closed before the streams were so protected. More details of how regulation on specific streams affected the hatcheries will be found in later chapters.

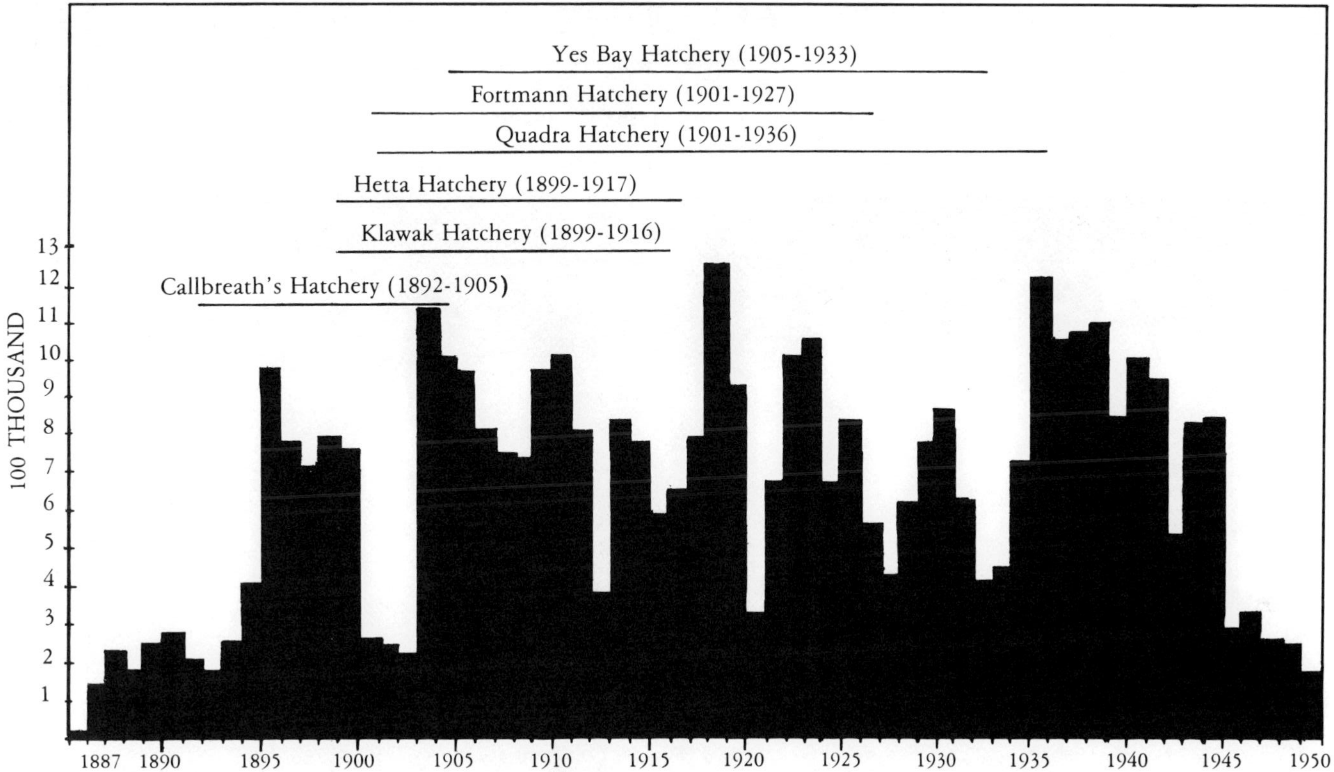

Figure 1. Commercial Catch, Sockeye Salmon, Southern Southeast Alaska in relation to Years of Hatchery Operations in that Area. (Source: *Larry Edfelt: Statistical History of Alaska Salmon Catches, Alaska Department of Fish and Game, Juneau, 1973.*)

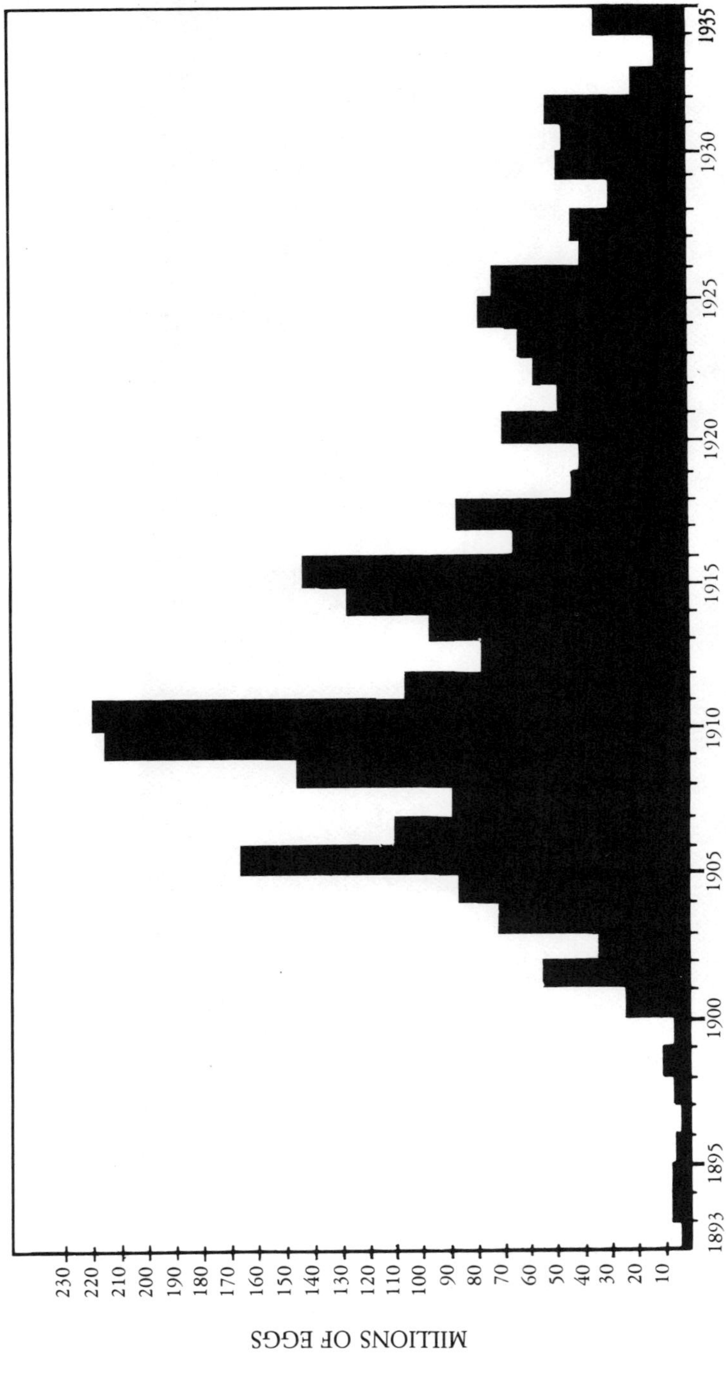

Figure 2. Number of Sockeye Salmon Eggs Taken, Hatcheries in Southern Southeast Alaska, 1893-1935.

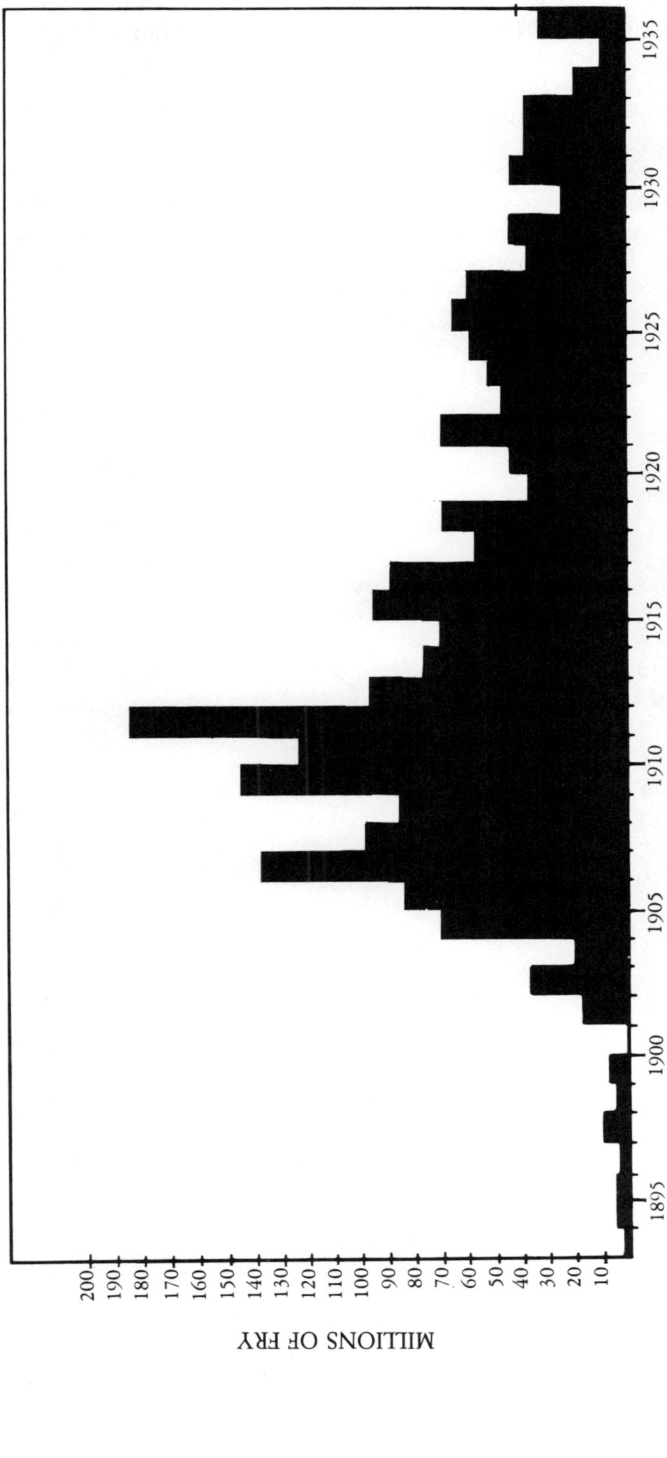

Figure 3. Number of Sockeye Fry/sac-fry Released by Hatcheries in Southern Southeast Alaska, 1893-1936.

The only hatcheries to produce fish by 1933 were the two federally operated hatcheries and one private hatchery at Quadra. That year a new United States Commissioner of Fisheries, Frank T. Bell, made a tour of Alaska and ordered the hatcheries to be closed. He announced that seven fish hatcheries in Maine, Vermont, Massachusetts, Ohio and Colorado would be closed. Bell took the view that the work of these hatcheries was a waste of public money and actually constituted an unjustified subsidy to a special industry, similar to the government planting an agricultural crop for private operators to harvest. On the other hand, he felt that a steady supply of fish would be provided by ample propagation. "The salmon will reproduce naturally," he said, "if a sufficient number are allowed to reach their spawning grounds. If we find that any district is threatened with depletion, we will restrict the gear or the fishing period and permit it to build up by natural propagation rather than try to do it artificially."[65]

A complete reversal of philosophy had taken place in the sixty years since artificial propagation was begun by the federal government in the United States. This policy of regulation instead of artificial propagation remained in effect in Alaska until recent times and there still are proponents of this policy.

Fish propagation was continued by one remaining private hatchery at Quadra after the last federal hatchery closed, but it, too, closed with barely any notice in 1936. Only one attempt to propagate salmon was made from that time until the 1950s. An experimental pink salmon hatchery at the U.S. Fish and Wildlife Service research station at Little Port Walter operated one year in 1947.

A renewed interest by the Territory in enhancement of its salmon resource occurred during the decade preceeding statehood. Although many of the projects of the newly created Department of Fisheries (1949) were not oriented to hatchery production of salmon, two hatcheries were built. The Deer Mountain hatchery in Ketchikan was designed to stock barren lakes. The one at Kitoi on Kodiak Island functioned as part of a research station whose main purpose was not to produce quantities of fish in a hatchery environment. Most of these salmon programs continued after Alaska gained control of its fisheries in 1960.

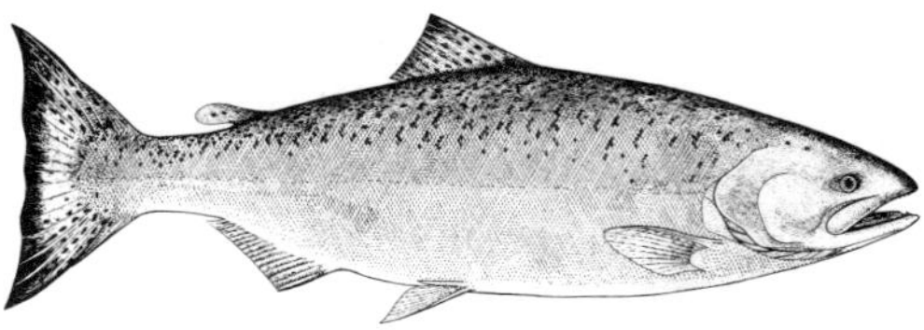

Notes

1. Livingston Stone, "The Artificial Propagation on the Pacific Coast of the United States with Notes on the Natural History of the Quinnat Salmon," *Bureau of the United States Fish Commission, Vol. XVI for 1896* (Washington: Government Printing Office, 1897).

2. Tarleton H. Bean, "Report on the Salmon and Salmon Rivers of Alaska with Notes on the Conditions, Methods and Needs of the Salmon Fisheries," *Bureau of the United States Fish Commission, Vol. IX for 1889* (Washington: Government Printing Office, 1891), pp. 207-208.

3. *Forest and Stream,* June 16, 1892.
Marshall McDonald, "Report on the Salmon Fisheries, *Bureau of the United States Fish Commission, Vol. XII for 1892* (Washington: Government Printing Office, 1893), pp. 14-19.

4. John N. Cobb, *Pacific Salmon Fisheries,* Bureau of Fisheries Document No. 839 (Washington: Government Printing Office, 1917), pp. 220, 243.

5. Unauthenticated recollections from various individuals in Southeast Alaska.

6. International Fishery Congress, *The United States Bureau of Fisheries: It's Establishment, Functions, Organizations, Resources, Operations and Achievements* (Washington: Government Printing Office, 1908).

7. Ibid.

8. Jefferson Moser, *The Salmon and Salmon Fisheries of Alaska* (Washington: Government Printing Office, 1899).
Jefferson Moser, *Alaska Salmon Investigations in 1900 and 1901* (Washington: Government Printing Office, 1902).

9. Howard M. Kutchin, *Report of the Special Agent for the Protection of the Salmon Fisheries,* Senate Document No. 153 (Washington: Government Printing Office, 1900), pp. 13-14.

10. Ibid, p. 16.

11. Letter from John J. Coyle to the Secretary of the Treasury dated 1902, U.S. Department of the Treasury, "Letters Received relating to Salmon Fisheries 4/20/89 to 12/20/97," Alaska File of the Secretary of the Treasury, 1868-1903, National Archives Microfilm Publication 720, Roll 19.

12. John J. Coyle, "Report of John J. Coyle, 1902, Assistant Agent for the Protection of the Salmon Fisheries in Alaska," Reports and Related Records, Division of Alaska Fisheries, Records of the U.S. Fish and Wildlife Service, Record Group 22, National Archives (hereafter abbreviated RG 22, NA.)

13. Ibid.

14. Howard M. Kutchin, *Report on the Salmon Fisheries in Alaska, 1900* (Washington: Government Printing Office, 1901), p. 14.

15. Jefferson Moser, "Report on Operations in Japan and Alaska Salmon Fisheries," Records concerning the Steamer *Albatross,* General Records, Records of the United States Fish Commission and Bureau of Fisheries, RG 22, NA.

16. Letter from the U.S. Commissioner of Fisheries George Bowers to the Acting Secretary of the Treasury, O. Spaulding, dated October 3, 1900, File "Cor respondence concerning Fisheries and Hatcheries in Alaska, 1901," Reports and Related Records, Division of Alaska Fisheries, RG 22, NA.

17. Kutchin, *Report on the Fisheries of Alaska, 1900,* p. 48.

18. Ibid.

19. Howard M. Kutchin, *Report on the Salmon Fisheries of Alaska-1903* (Washington: Government Printing Office, 1904.)

20. U.S. Congress, House, Committee on the Territories, *H.R. 9976, to Encourage Salmon Culture in Alaska,* Hearings, May 1, 1902 (Washington: Government Printing Office,) p. 7.

21. Ibid, p. 12.

22. Kutchin, *Report on the Salmon Fisheries in Alaska, 1900,* pp. 58-60.

23. Coyle, 1902.

24. *The Mining Journal,* [Ketchikan], May 11, 1901.

25. J.N. Wisner, "Alaska Report, 1903," File "Hatchery Inspection Report, 1903-1910, General Records, Division of Alaska Fisheries, RG 22, NA.

26. U.S. Congress, House, Message from the President of the United States, *Transmitting a Communication from the Secretary of Commerce and Labor Submitting a Preliminary Report of the Alaska Salmon Commission* (Washington: Government Printing Office, 1904).

27. Ibid.

28. Ibid.

29. U.S. Congress, Senate, Subcommittee on the Committee on Fisheries. *S. 5856, A Bill to Amend an Act for the Protection and Regulation of the Fisheries of Alaska,* Hearings, 62nd Congress, 2nd Session (Washington: Government Printing Office, 1912), p. 39.

30. Letter from Charles Nagel, Secretary to Honorable Attorney General, to James Wickersham, dated February 21, 1911, File "Afognak Island and Hatchery — Withdrawal and Reservation of Lands and Streams, 1903-1912," Reports and Related Records, Division of Alaska Fisheries, RG 22, NA.

31. Statement of Hon. James Wickersham, Delegate from Alaska, April 19, 1910, U.S. Congress, House, Committee on the Territories, *Amendment of Laws Relating to Fisheries and Other Occupations in Alaska,* Hearings, 60th Congress, 2nd Session (Washington: Government Printing Office, 1910), pp. 9-10.

32. Ibid, pp. 8-9.

33. U.S. Congress, House, Subcommittee on the Committee on Fisheries, *S. 5856* hearings showed that canneries paid $112,952 in taxes in 1911 and hatchery exemptions for that year were $42,642.

34. U.S. Congress, House, Subcommittee on the Committee on Fisheries, *S. 5856,* pp. 66-67. This number does not agree with official reports (Table 1). Using numbers in Table 1, rebates from 1906 to 1911 for Karluk and Fortmann hatcheries would be $203,762 — a $3,962 discrepancy.

35. Ibid, p. 68.

36. Ibid, pp. 68-69.

37. C.H. Buschmann, "Hatcheries and the Disposition of the Salmon Packers," *Pacific Fisherman,* November, 1914. The Northwestern Fisheries Company costs included operating costs, maintenance and ten percent depreciation. Cost per year at Hetta fluctuated greatly: 1909 — $296.99; 1910 — $499.97; 1911 — $390.14; 1912 — $967.10; 1913 — $878.08.

38. U.S. Congress, House, Subcommittee on the Committee on Fisheries, *S. 5856,* pp. 66-68.

39. No costs per million were found for feeding fry except by the Territorial Fish Commission. At the Ketchikan hatchery in 1926, costs were $210 per million fry from placing eggs in the troughs to release. The cost of pond feeding and pond construction was $810 per million for all species fed in fish and saltwater ponds. Eyak Lake hatchery in 1926 reported costs of $790 per million fry from green egg to release on sockeye salmon. (Alaska Territorial Fish Commission Report of the 1925-27 Biennium up to December 31, 1926, File "12-10-27", Alaska State Archives, Juneau.)

40. U.S. Congress, House, Subcommittee on the Committee on Fisheries, *S. 5856,* p. 69.

41. Ibid, p. 36.

42. Statement of Millard C. Marsh, May 2, 1910, U.S. Congress, House, Committee on the Territories, *Amendment of Laws Relating to Fisheries and Other Occupations in Alaska.* Hearings, 60th Congress, 2nd Session (Washington: Government Printing Office, 1910), p. 14.

43. U.S. Congress, House, Subcommittee on the Committee on Fisheries, *S. 5856,* p. 214.

44. Ibid, p. 232.

45. U.S. Congress, House, Committee on the Merchant Marine and Fisheries, 1916, *H.R. 9528, A Bill for the Protection and Regulation and Conservation of the Fisheries of Alaska,* Hearings, 64th Congress, 1st Session (Washington: Government Printing Office, 1916), Section 17.

46. Ibid.

47. U.S. Congress, House, Subcommittee on the Committee on Fisheries, *S. 5856,* p. 232.

48. Ibid.

49. Ibid.

50. Alaska, *Report of the Governor of Alaska to the Secretary of the Interior, 1917* (Washington: Government Printing Office, 1917), p. 33.

51. *Laws of Alaska, 1917,* "Chapter 75, An Act, S.B. 29," (Juneau: Territory of Alaska, 1917), p. 208.

52. Alaska, *Report of the Governor of Alaska to the Secretary of the Interior, 1917,* p. 33.

53. Territorial Fish Commission, File "12-10-1917", Records of the Office of the Territorial Governor, Alaska State Archives and Record Center, Juneau.

54. *Session Laws, 1919,* "Chapter 30, An Act. S.B. 10," (Juneau: Territory of Alaska, 1919), p. 176.

55. Alaska Territorial Fish Commission, *Report of the Alaska Territorial Fish Commission, Calendar Years 1923 and 1924* (Juneau: Alaska Daily Empire Print, 1924), p. 7.

56. *The Alaska Fisherman,* [Juneau] Vol. I, No. 2, April 11, 1923.

57. Ibid.

58. A.J. Sprague, *Report of the Territorial Fish Commissioner of Alaska to Alaska Territorial Legislature on H.C.R. No. 14, Session Laws 1927-28.* (Juneau), p. 4.

59. Ibid, pp. 5-6.

60. Compiled and computed from the annual Bureau of Fisheries reports *Alaska Fisheries and Fur Industries,* 1912-1959.

61. James A. Crutchfield and Giulio Pontecorvo, *The Pacific Salmon Fisheries: A Study of Irrational Conservation* (Resources for the Future, John Hopkins Press, 1969), p. 79.

62. Larry Edfelt, *Statistical History of Alaska Salmon Catches* (Juneau: Alaska Department of Fish and Game, 1973.)

63. Letter from E.A. Hitchcock, Secretary of Interior to Commissioner of General Land Office dated October 14, 1903. File "Withdrawal and reservation of lands and streams, also concerning Afognak Island and Hatchery, 1903-1912," Reports and Related Records, Division of Alaska Fisheries, RG 22, NA.

64. Letter from R.A. Ballinger, Secretary of Interior to Secretary of Commerce and Labor dated April 29, 1909. File "Withdrawal and reservation of lands and streams, also concerning Afognak Island and Hatchery, 1903-1912," Reports and Related Records, Division of Alaska Fisheries, RG 22, NA.

65. *Pacific Fisherman,* May 1936.

II. Procedures to Culture Salmon

Culture of salmon involves a series of methods designed to artificially produce juvenile fish that will mature in the natural environment of the ocean. This artificial production of juvenile fish can produce more fish than survive in nature.

It is necessary, of course, to understand the life cycles of the different species of salmon to succeed in artificially producing them. A brief description of the life cycle of each species of salmon is helpful to understand the complexities of artificial propagation. Chum salmon fry go into the estuary almost immediately after emerging from the gravel in the stream bed. From three to six years in the ocean is the usual time until maturity. Pink salmon go into the estuary, thence into the ocean, and mature after two years. The chinook salmon spend a year in freshwater then migrate to sea, reaching maturity in three to five years. Another species to spend a year in freshwater before migrating to sea is the coho. It matures in two to four years. The sockeye salmon normally spend a year in freshwater lakes downstream from the spawning area. The fish migrate to sea and mature chiefly at four years, but three to six year old mature sockeye salmon are found.

Initial artificial propagation of salmon in Alaska was influenced by limited knowledge of this natural history. This lack of knowledge was evident to scientists, but as F.M. Chamberlain concluded:

> ". . . few people care to work or pay for work on rivers and seas at finding out dull facts that can be brilliantly deduced from generalities in a laboratory. Salmon have nasty habits of choosing the very worst time of year for their purposes, getting into the most inaccessible places to breed, traveling on floods and doing other diabolical things to vex the observer. They even refuse to tarry long enough as youngsters to reach a marking age."[1]

The two salmon behaviors which concerned scientists most were the life history of each species of salmon and the "parent stream theory", that a salmon would return to the particular stream in which it was hatched. The two men most responsible for unraveling the mysteries of the sockeye salmon, the species most propagated in Alaskan hatcheries, were F.M. Chamberlain and Dr. Charles H. Gilbert. Chamberlain, in experiments conducted on the Naha River, in 1903, concluded that sockeye fry hatched at the Fortmann hatchery (on the Naha system) lingered a short time in the nursery ground where liberated, then moved down into the lake where they remained a year before going to sea.[2]

Dr. Charles H. Gilbert, who was a professor at Stanford University, and one of the country's foremost zoologists and ichthyologists and an authority on the life histories and habits of the Pacific salmon, developed a technique to read Pacific salmon scales.[3] By examining the concentric growth rings he was able to ascertain how long a young salmon stayed in fresh water before proceeding to sea, as well as the fish's age at maturity. Thus, Gilbert was able to show that the sockeye did, indeed, stay at least a year in a lake before migrating to sea, thereby confirming Chamberlain's observations.

It had been maintained by salmon fishermen, that mature salmon usually, if not invariably, returned to the stream where they were hatched. As early as 1906, Chamberlain announced as one of the conclusions of his study at Naha, Yes Bay and elsewhere in Alaska, that "at least the greater part of the supply of any stream must be derived from fry produced in that stream."[4]

Gilbert demonstrated the validity of the theory in experiments conducted in British Columbia around 1917 where he found that Fraser River fish returned to that river at their maturity and Gilbert concluded: ". . . this is true also of the fish of each of the large river basins. . . . It has now been shown . . . that this same principle holds in the case of all the rivers and creeks, however small these may be, and however near together they may enter the sea."[5]

Alaskan hatcheries started before the scientific work of Chamberlain and Gilbert so it was not known if sockeye fry stayed in fresh water, how long the fish stayed in the ocean to mature and if adult salmon returned to the parent stream. Many fish culturists and naturalists had theories. Many of them were later to be proven correct. It was with use of these theories, trial and error, and common sense that the first fish cultural procedures were developed in Alaska.

It was in 1897 that the U.S. Fish Commission published its first manual of fish culture to define methods used by the Commission to propagate all species of food fishes.[6] Although it was published under the authorship of John J. Brice, material was furnished by experienced fish culturists in the employ of the Commission. The chapter on Pacific salmon was provided by Livingston Stone, superintendent of the Baird, California, hatchery. Since chinook was the principal salmon that had been artificially propagated, the entire chapter was devoted to this species. It was, however, customary to ascribe to all species what was known of one. The introduction states "The methods described for hatching Salmonidae, while differing in minor particulars, are practically interchangeable, and may be used at the discretion of the fish culturist."[7]

This handbook, *A Manual of Fish Culture Based on the Methods of the United States Commission of Fish and Fisheries,* was available to Alaskans, but was soon replaced by a revised manual which came out in 1900, the same year that hatcheries became mandatory in Alaska.[8] Most of the salmon propagation procedures were taken verbatim from the earlier manual with added information from federal hatcheries. No mention of Alaska or sockeye culture was made.

The manual was not updated for 20 years, but the Bureau of Fisheries published many works on the subject even on propagation of Rhine salmon and on the Flodevig hatchery in Norway. Closer to home there were publications on food for young salmon, treatment of fish culture water for removal of algae, effects of disease on young salmon, treatment of fungus on fish in captivity, and methods for measuring eggs. There were 35 bulletins published in 1910 on various aspects of fish culture.

None related to the Alaskan experience, but some procedures were applicable. Ward T. Bower, a fisheries agent in Alaska, selected those procedures most applicable to Alaska and published them in the 1911 annual Alaska fisheries report.[9] He pointed out noteworthy advances such as taking eggs by abdominal incision and the use of the salt solution for removal of dead eggs. He wrote: "The value of these and other new methods has been thoroughly demonstrated, and it is only by adopting them that Pacific salmon culture can be brought to the new standard of efficiency."[10]

The next revision of the Bureau's fish culture manual, *Artificial Propagation of the Salmons of the Pacific Coast,* was published in 1920 and was written by Henry O'Malley, who, two years later, became the Commissioner of Fisheries.[11] This manual of fish culture referred specifically to Alaskan operations and techniques used. It updated the 20-year old manual and referred to results of investigations and study of salmon scales by Dr. Charles H. Gilbert.

Bower's 1911 report, O'Malley's 1920 Manual, and two editions of John Cobb's *Pacific Salmon Fisheries,* which liberally quoted the first two, were readily available and free to hatchery operators in Alaska.[12] It was probable they were even distributed without request upon publication.

In addition to these Bureau of Fisheries publications, fish culturists also had access to the *Transactions of the American Fisheries Society.* Up until about 1920, these proceedings contained many papers written by men who were experimenting in fish culture. Most Alaskan operators belonged to the American Fisheries Society during the years when their hatcheries operated, and thus received the published transactions.

Most Alaskan hatchery operators adapted artificial propagation techniques from the available literature to the uniqueness of their own situations. A.B. Alexander, U.S. Fish Commission employee, pointed out in 1897:

> "Alaska presents such a variety of conditions under which artificial propagation must be carried on, a plan which has been followed successfully in one locality may prove a singular failure in another. It is extremely doubtful if there can be found in the territory two places where the conditions are exactly alike and where propagation can be conducted in detail on precisely similar lines."[13]

This becomes evident in the histories of each Alaskan hatchery. However, some things are common, such as selection of a hatchery site, equipment for artificial propagation and the fish culture procedures.

SITE SELECTION

Criteria for selecting a site for a hatchery was rarely covered in the literature available in the earliest days of Alaskan fish culture. Selection was often hampered by the lack of knowledge of the terrain beyond the beach or stream areas. Jefferson Moser gave an example before a Senate hearing:

> "When I went to Loring in 1900 for the purpose of exploring Naha salmon streams there was not a white person who had ever gone beyond the lagoon [Roosevelt Lagoon]. They could not tell me anything about the interior of that country. Eventually I found an old Indian who claimed a hunting ground in the lakes, and with his assistance and his canoe, we managed to get at the upper lake system and to know a little something about it; but there were people who had lived around Loring for many years . . . and yet little was known of the interior."[14]

Sites were frequently selected on a hit or miss basis. Most missed. Of the major voluntary and mandatory hatcheries built prior to 1902, only the Fortmann hatchery on Heckman Lake remained on the original site chosen. Callbreath's McHenry Inlet, the Klawak, Hetta and Quadra hatcheries were all moved within the first year to more suitable locations. The Alaska Packers Association Karluk hatchery was built on another spot from the original facility. Most of the minor facilities built to satisfy the mandatory hatchery act failed because of poor site selection. Some washed away in high water; others had poor water or no run of adult salmon. The two federal hatcheries were placed after more comprehensive criteria for site selection were developed and both remained on the original site.

The first person to develop criteria for the site of a successful hatchery was J.N. Wisner, the fish culturist on the 1903 Alaska Salmon Commission expedition. Upon his return from Alaska, he put together a lengthy report, which, however, did not appear in published accounts. It contained criteria for selection of hatchery sites in Alaska:

> "It must be known that the fish still enter the stream in numbers that warrant a large hatchery. The fact they used to frequent it in numbers is insufficient. Every precaution should be taken against establishing a hatchery, provided only a few are erected, on a stream that would have to have its run of fish built up before it can supply the hatchery with all the eggs it can handle. . . .
>
> "The accessibility of the hatchery from the base of supply should be considered. Freight rates, added to the original cost of material, leads us to have reason to fear inaccessibility of hatcheries. . . .

"The building site will depend a great deal in connection with proper operation of the hatchery and it should receive more than a passing consideration before building is undertaken. . . ."[15]

Additional considerations Wisner felt should be taken into account were: possible use of waste water if building on a slope; need for a small amount of power to turn a grindstone, to operate a fish food grinder and to pump water for use in case of fire; and ease of putting in rearing ponds adjacent to the building so they could be worked without delay and to provide protection of the fish. He pointed out that buildings must be safe and be above the highest high water mark to prevent inundation.

One of the main considerations necessary for a hatchery location was the water supply. Wisner said:

"Water must be clear and pure and must have plenty of fall so that it may be brought into the building by gravity. If at all possible it should be a stream which has a spring as its source. If not, every attempt should be made to use a stream that is in the vicinity of spring water for the latter is invaluable against freezing. . . ."[16]

Almost every one of the early stations suffered the defeat of a year's operation when the water system froze. Open and even covered flumes could not conduct near-freezing water for any distance. As operators lived in the remote areas where the hatcheries were located, they soon discovered the only water which did not freeze was springs or spring-fed streams. These springs normally ran between 38 °F and 42 °F all winter.

There was one disadvantage to this warm water. O'Malley, in his 1920 fish culture manual, pointed out that spring water or spring-fed creek water was objectionable ". . . as it shortens the incubation period, bringing out the fry [sac-fry] at an earlier period than if hatched under natural conditions and at a season of the year when the natural food supply in the streams is at its lowest ebb."[17]

In Alaska heed was not paid to this observation, as there was still another advantage to spring water. Water taken from an open stream contained more sediment than spring water. Some of the vegetive matter tended to accumulate as slime among the eggs even to the point of smothering them. Those hatcheries which used stream water resorted to a variety of filters, the most elaborate at Karluk.

Conveyance of the water was almost always by gravity. Most common at first was to have water flow through V-shaped flumes above ground. Generally after one or two years of cold weather and freezing water, the water supply system was replaced by a pipeline, usually buried underground for protection. The intake to the system was most often located at a dam, which provided a reservoir of water for gravity feed into the hatchery. Often floods and anchor ice hampered the effectiveness of the intake systems. Screens were necessary to keep out debris.

Water flowed to the hatchery and into a headbox placed across the top of the rows of troughs. Once it had flowed through a row of troughs, the waste water was piped from the hatchery. At Karluk the waste water was used in a nursery pond. Waste water nursery ponds were also used for a brief time at the Yes Bay hatchery.

HATCHERY EQUIPMENT

All of the Alaskan hatcheries, except the Kitoi Research Station in the late 1950s, used what were known as Williamson troughs with wire baskets.[18] (Fig. 4) These were developed by John Williamson, of the California Acclimatization Society, with the aid of Marcellus G. Holton and Nelson Clark in 1873.[19]

The Williamson troughs were designed to provide upwelling water by use of permanent or movable partitions to regulate the direction of the water flow through the eggs suspended in the wire baskets. At first wooden and later metal partitions divided each trough into compartments slightly longer than the baskets to permit them to be raised and lowered and tilted. At the lower end of each compartment, a partition extended all the way across the trough and reached from the bottom almost to the top of the trough. Another similar partition at the upper end of the compartment reached from the top almost to the bottom of the trough. These partitions were about one to two inches apart. The water was consequently forced to flow under the upper dam, up through the basket of eggs and, thence, over the lower dam.

There was no uniformity in the dimensions of these troughs. Generally they were 16 feet long, but the hatchery at Quadra had a few six feet long, and the Hetta hatchery used 20 foot long troughs. The width varied from 14 to 16 inches and after 1904, the Bureau of Fisheries adopted a uniform measurement of 14 inches for trout and salmon hatcheries, whether private, state or government properties.[20] Depth of the troughs varied from Callbreath's five and one-half inches to seven and one-half inches at Quadra. The federal hatcheries used a six and one-half inch depth.

The troughs were arranged in pairs, two or three pairs placed end to end on different levels, with a fall of water between each pair. The length of this fall varied from hatchery to hatchery. Brice's 1897 fish culture manual suggests one and one-half inches.[21] O'Malley's 1920 manual suggests about six inches between pairs.[22] The Alaskan federal hatcheries used one and one-half inches. Klawak and Callbreath's hatcheries used four inches. The troughs were provided with covers made of canvas stretched upon light frames and made sunproof by saturation with asphaltum varnish, which had also been thickly applied to the inner surfaces and to the egg receptacles. These covers were used to exclude light during certain stages of incubation.

Figure 4. Williamson troughs with wire baskets removed from trough, Karluk hatchery, 1914. *Bureau of Fisheries.*

The egg receptacles were oblong-mesh wirecloth baskets which were about 24 inches long, and varied in width depending upon the width of the trough. They were deep enough to project an inch or two above the water. Some baskets had wooden rims which rested on the top of the troughs and held the basket an inch off the bottom. If there was no wooden rim, other devices were constructed to keep the basket suspended above the bottom. An angle iron along the length was one method. Sometimes a false bottom of wirecloth or perforated metal was placed about an inch to one and one-half inches from the real bottom. Any sediments would fall below the false bottom aided by the movement of the fish.

A metal rim was adopted just prior to 1920, and it rested on the top of the trough, supported by metal legs fastened to the edge of the basket. The basket was kept flush with the lower riffle dam to force the ascent of the water through the basket. The number of eggs that could be incubated in a basket depended upon the species of salmon. It was customary to place about 50,000 sockeye eggs per basket. O'Malley's 1920 manual gives the number per basket at 20,000 to 30,000 for chinook, 50,000 to 60,000 for sockeye, 30,000 to 35,000 for coho, 40,000 to 50,000 for pink and 33,000 to 38,000 for chum.[23]

Williamson troughs were not only used for incubation of eggs, but also for holding some of the sac-fry to the swim-up stage. As the first eggs began to hatch, the sac-fry in the baskets were removed and either placed in ponds or released. When all eggs had hatched, the baskets and riffle dams were removed from the troughs and each trough was divided by screens into three compartments. The most recently hatched sac-fry were placed into the troughs, where they were held until they developed into fry. At some hatcheries fry were fed in troughs and released prior to taking of next year's spawn.

ADULT CAPTURE AND TAKING OF SPAWN

The spawn (eggs and milt) used for artificial propagation was obtained from salmon taken on their upstream migration to the natural spawning grounds. At Karluk adults were held in saltwater corrals and later in freshwater ponds until the fish were ready to spawn. Large losses occurred each year. The only other facility where sockeye were held was at the Territorial Fish Commission hatchery at Eyak Lake. A web corral at the outlet of the lake held the fish, but they fought the barricade causing injury and exhaustion. Five hundred were lost in one week and after 1,380 died, it was obvious that attempts to hold the sockeye adults were futile and the remaining fish were liberated.

The most successful means of capture was to build a picket barricade across the stream, preferably near its mouth. This barricade was called a rack, or what is known today as a weir. Descriptions of various styles of racks are in all early fish culture manuals. Some were built of wooden slats placed on pile driven aprons such as the one at Yes Bay hatchery (Fig. 5). Quadra used one which utilized iron rods, in a similar fashion to many Alaskan weirs today. The rack stopped the passage of the fish at a spot where the haul seine could be effectively utilized. Large trees which accumulated during the year off the mouth of tributaries from floods were cleared away before seining began. Then, as the fish schooled below the rack, a drag or haul seine was used (Figs. 6 and 7). At the Yes Bay hatchery, the net was attached to the rack and then maneuvered around the fish. The least effective way to capture fish was used at Hetta and Fortmann hatcheries where a haul seine was used without benefit of racks.

Two stations used traps installed in the stream which allowed the fish to swim in but not out. Callbreath used traps in most of the small tributaries to his lake, and Quadra hatchery used an auxiliary trap shown in Figure 8.

Once the salmon were captured, it was necessary to sort them to select those which were mature or ripe enough to spawn. Of the signs which usually accompanied ripeness in a female salmon, the separation of the eggs in the ovaries was the surest. But the early fish culturists, according to Brice in 1897, relied upon "an indescribable ripe look, which is neither color, shape nor conditions of the organs, but a general appearance which shows at a glance that the fish is ripe and can be appreciated only by experience."[24]

Figure 5. Rack or weir at Yes Bay hatchery, 1929. *Courtesy of A. Kemmerich.*

Figure 6. Capture of spawners at base of weir, Afognak hatchery, 1914. *Bureau of Fisheries.*

Figure 7. Capture of spawners with haul seine, Yes Bay hatchery, 1923. *Bureau of Fisheries.*

Green fish, or fish which were not ready to spawn, were returned to the lake and there were many days at Quadra hatchery when as many as half the fish were green. Ripe fish were segregated by sex and placed in pens or live boxes. Fish were spawned nearly every day as it was believed the eggs of the female confined in pens were often injured. The usual procedure at the turn of the century was to take the fish from the pens in dipnets and hold them suspended until they became quiet enough to handle. During the early hatchery programs, fish were rarely killed to obtain spawn. The objection to killing fish before spawning was the large loss of eggs resulting from killing partially ripe fish.[25] For larger fish, the man taking the spawn, held the female or male in one hand while another man held the tail. The eggs or milt were then pressed from the fish. Smaller fish were spawned by one individual. (Fig. 9) Sometimes a finger was inserted into the vent, enlarging it to enable an easier flow of eggs.

This method of taking salmon spawn was practiced in Alaska for many years. It was replaced by the abdominal incision method in 1911 at some of the hatcheries, but many kept the old practice until forced to abandon it by the Bureau of Fisheries.

Figure 8. Instream trap for capture of spawners at Quadra hatchery, circa 1935. *Courtesy of Clare Ritter.*

The method of taking eggs by incision, basically what is used today, was discovered and developed around 1903 by Cloudsley Rutter in connection with a study of the life history of salmon on the Sacramento River. Fish were first killed by a blow on the back of the head and some culturists then bled the fish, usually by cutting off the head or tail. Bower in 1911 expressed the opinion that experimentation had conclusively demonstrated that no advantage resulted from the bleeding procedure and meant only extra labor.[26] By 1920 O'Malley advocated bleeding of fish.[27] Personal choice among fish culturists seems to have played a major role, and a former U.S. Fish & Wildlife Service employee feels he still can take eggs more quickly, with as little blood on the eggs, as anyone who takes time to bleed the fish.[28]

In taking eggs by the incision method a cut was made in the abdominal walls from the throat or near the pectoral fins to the vent. When the cut was made, the knife was either shielded by a guard or held betwen thumb and forefinger so that less than a half inch of the blade projected. This precluded the possibility of injuring the eggs.

Two different methods were used to bring the milt and the eggs together, the wet and the dry methods. In the wet method, a little water was considered desirable "to give greater activity to the milt" as Brice described it.[29] The water was placed in the pan, the eggs stripped into it and then the milt. Brice continues "If left more than a minute in water there is a decided loss of fertilizing power. . . .It is not prudent to leave the eggs in the water a moment longer than is necessary before adding milt."[30] The milt was known to retain its fertilizing power several days when kept from air and water.

The addition of water was not essential to good impregnation, others believed, so they used the dry method. This method, apparently first discovered by a pioneer fish culturist, Seth Green, in 1864 for use with trout, consisted of taking both eggs and milt in a dry pan. While considered the first great advancement in fish culture, this method was still not always practiced in the earliest years in Alaska. By 1920, O'Malley discusses only the dry method.[31]

Various procedures were in use after the eggs were impregnated. The spawn might be left in the pan until it separated, when it was then washed clean and placed in troughs. Another technique was to pour the contents of several pans into a large can immediately after impregnation. When the eggs separated, the contents of the can were poured into the hatchery troughs, where the current washed away the milt. Where water supplies were scant and current sluggish, the milt was thoroughly washed from the eggs before placing them in troughs. Washing milt from the eggs was standard by 1920.

The procedure of water hardening eggs first appears in the literature available to Alaskans in 1912.[32] There is a sensitive period occurring about an hour after impregnation when eggs should not be handled. Many Alaskan operators placed eggs in the baskets and troughs before this period of time. If

Figure 9. Taking of spawn without an abdominal incision, Klawak hatchery, 1910. *Cobb collection, University of Washington.*

not, they were left in buckets until water hardened or ''separated'', as many fish culturists called it.

Methods of enumerating the eggs at each hatchery varied. At first a standard fecundity (number of eggs per fish) of 3,000 to 3,500 sockeye eggs was used. Another method was that of measuring eggs by a dipper or bucket of known average egg capacity. Callbreath used an Eagle Brand condensed milk can (1,900 sockeye eggs and 850 coho eggs).

When it became well established that fish eggs increased in size during periods of incubation, the Bureau of Fisheries suggested that to avoid error through this change, all standards of computation should be determined by actual count. But many of the methods designed to calculate exact count through displacement were considered by Bower in 1911 to be less desirable in their practical applications than the common dipper or bucket method.[33] With no general agreement on best procedures, no standardization of egg enumeration was ever accomplished. During the period of hatchery rebates, this created a problem.

INCUBATION

Early fish culturists in Alaska knew that the period of incubation of salmon eggs depended upon the water temperature. The Karluk hatchery in 1896 heated the incubation water in an attempt to hasten the incubation period. This was the only hatchery in Alaska which actually did this, although some of the smaller Southeast hatcheries mentioned the desire to do so.

Temperature units were used at some of the hatcheries to calculate the rate of growth during the incubation period. A description of this method was first published in 1901 and was available to hatchery operators.[34] A temperature unit is the daily average water temperature. By summing the daily temperature units (T.U.) the occurrence of various life stages in salmon development could be predicted.

For those not using temperature units, O'Malley had what he called a safe rule, originated by Seth Green: "In a temperature of 50 °F the eggs will hatch in 50 days. Every additional degree of warmth lessens the incubation period by 5 days and every degree lower than 50 prolongs it 5 days."[35]

Little information on incubation periods and required temperature units is available. Incubation record keeping was not standard and was only a suggested requirement.

At the Fortmann hatchery it was the custom to remove dead eggs on the day after fertilization, then to leave the eggs undisturbed until the spinal column was well formed. At this time, eggs were considered hardy enough to be packed for shipment or to have dead eggs removed. Both Brice and O'Malley in their fish culture manuals refer to the spinal column formation rather than the first appearance of pigmented eyes. However, the term "eyed egg" appears in much of the early literature.

Other fish culturists often found that eggs became buried under or in a mass of mud and sediment, so it became the practice in many hatcheries to pick the eggs throughout the entire period of incubation. Eggs were handled by skilled operators with utmost caution at all times except during the two or three days prior to the time a perceptible curve showed in the tail. By picking daily there was a loss of between three to five percent, but the loss from fungus was thought to be greater.[36] Picking was done with tin forceps, and a feather was used to stir the water which gently brought the eggs to the surface. At many of the Alaskan stations this method was used exclusively to remove dead eggs.

Another method, considered at the time to be one of the more noteworthy advances in fish culture methods, was the use of salt as an aid in removing dead eggs. This idea was first presented in 1905 to the American Fisheries Society by Henry O'Malley, then a culturist at the Baker, Washington, hatchery.[37] The refinement of his experiment was carried on at the Yes Bay hatchery first by Wm. Hancock, superintendent and later by L.E. Baldridge. It became, by 1911, a part of the Bureau of Fisheries recommended procedures of fish culture.

The principle of the salt solution is that specific gravity of good eggs is greater than that of bad eggs, hence, upon being placed in salt solution, the good eggs sink and the bad remain afloat and can easily be removed. It was vitally essential that the solution was of the proper strength. When too weak, all eggs sank, while if too strong, all floated. The margin of the proper density was so narrow that during the proceedings it was necessary every few minutes to strengthen the solution, as the small amount of fresh water adhering to the basket of eggs as it was lifted from the hatching troughs into the solution, affected the strength. The normal solution consisted of one part salt to nine parts water.

A detailed description is given of the methods used at the Yes Bay hatchery:

> "The chief item of equipment consists of a water-tight wooden tank, 4 feet long, 2-1/2 feet wide, and 10 inches deep, for holding the solution in which the eggs are immersed. . . .It is necessary that the surface of the solution be perfectly quiet, for any ripple or current will tend to disturb the bouyant effect of the solution upon the eggs. Therefore it was found of great convenience. . . to use a floating frame made of half-inch material 6 inches wide fastened together vertically and at right angles, thus forming open squares about 6 inches either way. After each basket of eggs is lifted from the salt bath this frame is placed in the solution to stop all motion of the water, being pushed down until it is almost submerged and held firmly against the side of the tank for a few seconds. . . .
>
> "Another piece of equipment is a feather fan with which gently to push the floating dead eggs away from over the submerged basket into which the good eggs have settled. Unless the dead eggs are quickly moved, they too will sink. A feather fan made of fastening eagle feathers to a thin strip 8 inches long by 1-½ inches wide works much more satisfactorily for this purpose than a wing. An ordinary hand scaff net about 12 by 14 inches in size for removing dead eggs from the tank, a dipper and a bucket complete the outfit."[38]

The salt solution procedure was used as soon as the fish and distinct curvature of the tail showed plainly when the egg was held to the light. The eggs were then stirred thoroughly which caused the unfertile eggs to turn white. After this, the baskets were returned to the troughs and allowed to remain three days, at which time the salt solution was applied. Two men were necessary for this procedure, one to dip the basket and the other to feather the eggs. The bad eggs on the surface of the brine were immediately removed and after five or six baskets were treated, all dead eggs which had sunk to the bottom of the tank were removed so as not to weaken the solution. If failure to treat a basket occurred from a sudden jar or from brine too weak or too strong, the basket was returned to the trough as it would not respond to treatment again the same day. When O'Malley described the procedure in 1920, a second box provided with handles and screen bottom into which the basket of eggs was placed had been added.[39]

The main advantage of the salt solution over the time-honored process of picking by hand was obviously a great savings in labor. It was established that if the eggs were of not more than mediocre quality, it would take at least 20 pickers to remove as many eggs as could be done by two men using the salt solution.

Another advantage was that it was possible to clean the eggs more thoroughly. This greatly reduced loss from contamination and infection resulting from decomposition and fungus growths which inevitably followed the long presence of dead eggs. The early hatchery operators picked their eggs daily after the initial resting period, if the salt solution was not used, to eliminate this possibility.

In addition to the Yes Bay station where the majority of the work on this technique was developed, the Afognak station used the salt solution method. The Fortmann hatchery began its use in 1911, the Hetta hatchery used it the last few years of its existence, and the Quadra hatchery was using it by 1923. It was never used at Karluk and it is not known if the Territorial hatcheries used it.

CARE OF SAC-FRY

Nothing has caused more controversy over the effectiveness of the early hatchery programs than the methods used to care for and release the hatched fish. The most publicized and best known of the early hatcheries was at Karluk. Much of the criticism of salmon hatcheries in Alaska has been made from assumptions that all hatcheries functioned in the same manner: Karluk used techniques proved to be ineffective; therefore, all Alaskan hatcheries used the same techniques and were ineffective. Karluk's method of treating sac-fry and fry was so unusual and atypical of Alaska that most of the details of its operations will be confined to the station history.

Once the eggs hatched, there was no standard procedure developed by the Commission of Fish and Fisheries or by the Bureau of Fisheries for sockeye salmon. It appears there were as many theories on the correct way to care for the resultant fish as there were fish culturists and naturalists. Most people actively propagating Alaskan sockeye salmon learned by trial and error. Available space, money and pre-existing equipment dictated whether a new theory was tried.

The sac-fry were released within several days of hatching during the early years. This was necessary because of the small space available in most hatcheries. These sac-fry were planted in the home or hatchery stream, which was a tributary to a lake or were planted in the lake or along its shores. Karluk was the only hatchery which planted fish in brackish or salt water. The Karluk hatchery was located at the mouth of the Karluk River, on a lagoon, about 16 miles from Karluk Lake. The site was selected before it was proven sockeye fry spent a year in fresh water.

Research between 1900 and 1910 at various Bureau of Fisheries hatcheries on the Pacific Coast showed the desirability of holding sac-fry until the sac was nearly absorbed. Bower, in his annual Alaska fisheries report in 1911, brought this to the attention of Alaskan hatchery operators. He wrote:

> ''No matter how thorough the work in the hatchery proper may be, no adequate or justifiable return can be expected unless every possible precaution is likewise taken to conclude the process properly. There is no justification whatever for artificial or, perhaps, as it may be termed 'protected' propagation where the returns to the water are not made in the most scientific and intelligent manner possible. Otherwise it were better to let nature do the work.
>
> ''Especially is there lack of consideration as to the proper age at which to plant young fish, and again as to the selection of places for releasing them in order to give the best results.''[40]

Bower believed the best results would be obtained when young sockeye were planted with the umbilical sac two-thirds absorbed, which he said was the time fry began to swim up freely. ''With the temperature prevailing at the Alaskan hatchery this means that the fry must be held at least four or five weeks after hatching.''[41]

The holding capacity in troughs for young fish was much greater than that necessary for eggs. Some hatcheries such as Yes Bay and Afognak put in extra troughs. But most hatcheries reduced the number of fish in the troughs by removing the oldest sac-fry as the eggs hatched. After all eggs were hatched and the baskets and riffle dams were removed from the troughs, each trough was divided by screens into three equal compartments for holding a certain portion of fry. The number of sac-fry which could be held was about 50,000 per trough.

A method for increasing the capacity to hold the sac-fry in the hatchery trough was the stacked tray system. It was developed and tested by Joe Kemmerich at the Bureau's hatchery at Baker, Washington, around 1920. The system was adopted by both federal hatcheries in Alaska, but none of the private hatcheries utilized this method.

With the stacked tray system, the eggs were carried in the regular baskets in the troughs until within two to three weeks of the hatching period. At that time they were transferred to 14 by 16 inch trays made of 1 by 1.5 inch lumber, covered with wirecloth 14 to 16 meshes to the inch. These were treated yearly with asphaltum varnish. From 3,500 to 4,000 sockeye eggs were placed on one of these trays. Then the trays were set one on top of each other, six to eight deep in the trough compartment, and a lid placed on top. The trays were clamped together and wedged firmly in the trough. A strip, ¾ inch, was placed around the bottom tray thus forcing the water to upwell through the stack. With the stacked trays, the holding capacity of a 16 foot trough was increased from six to eight times and permitted the safe development of 300,000 to 400,000 fry in a space formerly required for 50,000.

It was possible to seed green eggs onto the trays, but it was found to be more satisfactory to eye the eggs in baskets and remove dead eggs before use of the trays. This permitted a thorough cleaning of all apparatus involved. On the trays, egg shells left by hatching sac-fry were ground up by movement and soon floated free. This helped prevent fungal growth.

During this period of incubation on the trays, the sac-fry developed under conditions which were believed to be characteristics of gravel-incubated fry.[42] The sac-fry were more docile and did not utilize the food in the yolk-sac for movement, as they did in open trough systems where sac-fry tended to constantly move to and pile up at the end of the trough. To keep them from smothering in open troughs, it was necessary for hatchery personnel to "paddle" or "stir" the sac-fry over the full length of the trough compartment by means of a rolling of the water with a paddle. The fish were not touched, but were moved by the gentle agitation of the water. This was done twice a day. The process of "paddling" was eliminated by the tray system.

The tray system proved to be highly successful and mortality rates were low. Alphonse Kemmerich, a superintendent at the hatchery at Yes Bay, remembered that in six to eight weeks the death rate was as little as five to six fish per tray and these were usually deformed.[43]

C.H. Van Atta, another superintendent at the Yes Bay hatchery wrote of the tray system:

> "Sixty thousand sockeye salmon fry were placed on 12 trays, each 13-½ by 25 inches in dimensions. The trays were stacked in one regular basket compartment of the hatching trough, and held thus from March 12 until the yolk sac had been completely absorbed, on June 7. The total loss of fry during this period was 180."[44]

The only recorded problem with the tray system was reported in 1922. The superintendent at Yes Bay hatchery at that time wrote that during that year a loss in excess of 29 percent occurred among the fish thus held, while the death rate among the sac-fry held in troughs was normal.[45] He attributed the abnormal loss to an unusual condition of the water supply, not reported previously. Circulation almost entirely ceased at times by the presence of quantities of slime which covered the frames and wires of the trays, forming a jellylike substance. This slime was supposed to have come from the swampy surrounding area. No other such occurrence was recorded.

Once the fish were swimming, many Alaskan hatcheries released the fish in the hatchery stream or along the lake shores. Other operators used ponds in which fry were held in an attempt to increase their growth before release. The first two to use rearing ponds were the Karluk and Callbreath's operations, both prior to 1900. Robert D. Hume, a friend of Callbreath's, who built and operated a hatchery on the Rogue River in Oregon, was the first to take up rearing salmon fry on any scale, and undoubtedly influenced Callbreath to release sac-fry in springs which were dammed. The Fortmann hatchery put in its first sac-fry rearing ponds in 1903 and Quadra in 1908.

The Alaskan rearing ponds varied at each hatchery and often from year to year at the same hatchery. The Quadra hatchery developed a series of ponds within two tributary streams of the main inlet to Hugh Smith Lake. The Yes Bay hatchery experimented with this same method. At other hatcheries ponds were dug and lined with gravel and boulders to simulate the natural environment. (Fig. 10) Some were too shallow and froze solid, but the Territorial Fish Commission reared sockeye and chinook to the fingerling stage in hatchery ponds. Operators at Hetta hatchery tried a wire corral in the lake off the mouth of the hatchery stream. At Yes Bay hatchery a slough or arm of McDonald Lake was fenced for a large rearing area. This method was used at both the Eyak Lake and Grouse Lake hatcheries of the Territorial Fish Commission in the 1920s.

The Territorial Fish Commission at its Ketchikan hatchery utilized rearing ponds in still another manner. Pink and chum salmon were reared in salt water in enclosures built in small bays. Callbreath in 1895-96 and the Territorial Fish Commission in 1919-20 experimented with using barren lakes as nurseries: Callbreath used sockeye salmon and the Commission used coho and pink salmon. In the 1950s the Kitoi Research Station and Ketchikan's Deer Mountain hatchery began scientific refinement of barren lake stocking.

FEEDING OF FRY

Very early, the fish culturists were aware of the benefits of feeding the fry to build up their strength "to better enable them to cope with their natural enemies or other unfavorable conditions" as Bower put it.[46] Early fish culture manuals all agreed that feeding of salmon should begin as soon as the fish were capable of swimming up, and this was the common practice in Alaska.

Figure 10. Fry rearing ponds, Fortmann hatchery, 1910. *Cobb collection, University of Washington.*

However, the greatest handicap to artificial rearing of salmon fry was the lack of a satisfactory food supply. The food had to be cheap, easily kept, and available in large quantities. Often it was found that food which met these criteria was unacceptable to the small fish.

The first Alaskan hatchery operator to attempt feeding fry was James A. Richardson at Karluk in 1900. Callbreath did not feed his fry except an occasional bit of dried salmon if they looked hungry. A few other hatchery operators during the period between 1905 and 1910 tried an occasional experiment to determine what food the fry would eat and, in most cases, would not eat.

Discouraged at the poor results, most of the private hatcheries did not attempt to feed their fry. At Hetta, Quadra, Klawak and Karluk little, if any, feeding was done. Fortmann was the exception. In the earlier years, the operators were so interested in providing food for their fry, they attempted to grow plankton in trenches connected with the lake, but in later years interest waned. The Yes Bay station and the Territorial Fish Commission hatcheries were the most successful in their fry rearing programs.

The foods most often fed in the earliest days were herring roe, canned salmon (do-overs or cans not properly sealed and thus condemned for human consumption), salt salmon and fresh salmon or trout. At the federal hatcheries, the carcasses of the spawned salmon were salted in barrels, and later freshened and prepared. Other fish captured in gillnets and traps were prepared fresh. All of these foods met with varying degrees of success, except for herring roe which proved to be unsuccessful.

Figure 11. Hand grinder used to prepare fish food, Yes Bay hatchery. *Cobb collection, University of Washington.*

Condemned liver was considered one of the best foods, but was never cheap nor easy to transport without spoiling. It was generally cooked and ground, particularly after 1915 when two scientists at Reed College in Oregon published a paper which showed by their experiments fish fed cooked liver gained weight faster than those fed raw liver.[47] Salmon eggs and milt, as fry food, gained popularity in the 1920s after a paper was read on the subject at the American Fisheries Society meeting.[48]

Any food had to be prepared small enough to be utilized by the fish. At first hand grinders were used (Fig. 11). Later grinders were connected to the power system. At most hatcheries the fry were fed until fall when spawn from a new brood was taken. At that time they were released into the lake, often by opening pond gates to allow fry to disperse volitionally. Occasionally fry chose to overwinter in the ponds, but were not fed. Amounts of food, time of feeding and length of feeding time varied from hatchery to hatchery. The Ketchikan Territorial hatchery experimented with feeding fry over winter, but that was the exception.

DISEASE

Every fish culturist knew that not all of the eggs taken would survive to hatch, and they knew, too, that sac-fry and fry could become diseased and die. Since little was known about curing diseases, prevention was emphasized. Fungus, or byssus as it was first called, grew in wooden troughs and many eggs were lost unless the troughs were kept scrupulously clean. The first U.S. culturists used charred wood to construct troughs and to eliminate the fungus problem. Later Seth Green used asphaltum paint to prevent fungus.[49] Removal of dead eggs was considered essential to fungal prevention. There are very few reports of loss of large numbers of eggs from fungus during the incubation period. Exceptions were the original hatcheries at Karluk and Klawak, both of which lost the majority of their eggs in the first year.

A few other incubation problems were reported. One such problem was at Yes Bay in 1923 when about 15 percent of the eggs were lost. The water became heavily charged with gas, which formed masses of tiny bubbles on the baskets and upon the eggs. The constant bursting of the bubbles agitated the eggs, causing them to accumulate at the lower ends of the basket. The cause of the abnormal condition of the water supply was not definitely determined but it disappeared after fall rains and cooler weather began.[50]

At first no fry diseases were reported because nearly all fish were planted as sac-fry in lakes, streams and, later, rearing ponds. What little can be learned of fry disease problems facing the early fish culturists comes from reports of the Bureau of Fisheries.

The earliest reported incidents of disease in sockeye fry occurred at Afognak and Karluk. Sockeye eggs taken at the Uganik Bay field station in 1915, developed white spots in the egg about 10 to 15 days before hatching. The sac-fry also had a white spot and died in the fifth or sixth day. This had never occurred at Afognak before. At Karluk in 1916, the Bureau of Fisheries Inspector, Ball, found the same symptoms in the sockeye sac-fry. He writes:

> "A close examination of the fry . . . disclosed the fact that one-tenth of them were afflicted with an internal injury or disease manifested in the form of a white spot or blotch on the umbilical sac. Usually near the vital organs. At first, it was thought that this might be a condition of the skin only, originating after hatching, but subsequent examination confirms the belief that it is a deep-seated malady, one that is fatal and that it has reached a stage of noticeable development during incubation."[51]

The other recorded instance of disease at Afognak occurred in 1921 and 1922. Few details survived of the symptoms. Sac-fry appeared normal at the time of hatching, and the trouble appeared just prior to the feeding stage. In May about 12 million fry were attacked by the disease and great mortality was experienced. The disease did not respond to any of the methods of treatment known at that time. The surviving fry were planted in Litnik Lake and shortly after this the yearling salmon in the lake appeared to be affected by the same malady, perishing in large numbers. Samples of both the fry and the yearlings were sent to the Division of Scientific Inquiry, but records of the analysis were not located.

The best documentation of symptoms during an epidemic was made at the Yes Bay hatchery in 1933 by Vincent Boucher, foreman who kept the station logbook. During February, an abnormal death rate was observed in three troughs. Fish were put in hospital troughs and dead removed. Some days later fish in various parts of the hatchery showed apparent signs of distress. The stacked trays were removed, brushed and troughs cleaned. Soon, it was apparent an epidemic was in progress.

Boucher recorded the following symptoms:

"1) Faded appearance of dark colored portions of body.
2) Great distention of yolk sac.
3) Apparent rupture of veins in yolk sac.
4) Presence of gas bubbles in yolk sac, many of these being as large as 1.5 mm in diameter.
5) Presence of oil globules of similar dimensions in yolk sac.
6) Aimless and distress swimming of fry on surface.
7) Inability of fry to leave the surface.
8) Floating of fry in supine position."[52]

Many attempts were made to determine the cause of the epidemic. The only apparent source of introduced infection from other than local sources was a shipment of eastern brook eggs received from Oregon. It was noticed that the signs of infection were first observed in the part of the hatchery in which the egg transport cases were emptied of their contents. Boucher said:

> "Yet this would scarcely seem to explain the infection of the entire line of three troughs since neither the case nor the eggs came in contact with troughs other than the lowest of the battery. Nor would it seem to explain the finding, in mid-March of one tray (the middle one), in the stack farthest from the foot of the battery, in which no single fish remained alive; and in which all had, very evidently been dead for weeks."[53]

When it was discovered that the naturally spawned fry coming downriver were dead, Kemmerich, then superintendent, sent Boucher and others to search the water source to see if they could find anything unusual. This attempt met with failure as the underbrush and terrain was not conducive to such a search. Kemmerich observed that later the fry and fingerlings in the lake began to perish also from the same affliction.[54] Samples were again taken, but the station was closed before results were obtained.

Few records of diseased fish were found in connection with the private hatcheries. Territorial Fish Commission records are so meager that no problems with, or treatments for, disease were discovered. In summary, it was found that the literature on Alaska hatcheries dealt little with the problems of and solutions to salmon diseases.

EXPERIMENTAL WORK

Experimental work was conducted to provide alternatives to the pattern of taking spawn, incubating eggs in Williamson troughs and rearing sac-fry and fry for release. Three alternatives were explored: planting of green or eyed eggs (Table 2), utilizing barren lakes and using gravel in containers to incubate eggs.

Generally fish culturists in Alaska did not choose to plant freshly fertilized salmon eggs (or green eggs). This usually occurred as a consequence of an abnormal situation. Callbreath planted green sockeye eggs in 1901 at Salmon Bay after arrangements for transportation were not successful. During the years the Yes Bay station operated field stations, it was necessary twice to plant green eggs. When collecting pink salmon spawn at Ketchikan Creek, the citizens of the town protested and, to appease them, the collected eggs were planted in the stream gravel. The pink eggs taken at Smeaton Bay were so few, they were planted in the gravel of streams in the area. Only A.J. Sprague, during his term as hatchery superintendent of the Territorial Fish Commission, planted green eggs indiscriminately, moving green eggs and planting them to perpetuate new runs.

Eyed eggs were planted more often in Alaska. Its first use came when normal hatchery procedures were interrupted. This occurred most often at the turn of the century when the water supply froze or storms carried away the supply flume or pipe. Once these problems were solved, the privately operated hatcheries did not plant eyed eggs. Eyed eggs were planted on a few occasions at Afognak and in the experiments on barren lakes by the Yes Bay station.

It was A.J. Sprague, hatchery superintendent of the Territorial Fish Commission, who believed in and planted the most eyed eggs. He believed that the advantage of eyeing eggs and subsequent planting of them were:

> "(a) saving one-half of the cost of plant, equipment and operating costs;
> (b) eliminating the entire cost of retaining ponds and feeding of the fry;
> (c) preventing the loss of fry in permitting them to obtain natural food in their own way. . .
> (d) providing the retention of the natural instinct of self-preservation."[55]

Table 2. Miscellaneous Eyed and Green Salmon Egg Distributions, Alaska, 1901-1953

Year	Species	Number of Eggs	Source	Disposition
1901	sockeye	...	Salmon Bay	Planted green, Salmon Bay Creek by J.C. Callbreath.
1903	sockeye	600,000	Buschmann Creek	Planted eyed by Quadra hatchery.
1905	sockeye	2,000,000	Klawak Lake	Planted eyed along lake shores by Klawak hatchery.
1906	sockeye	...	Klawak Lake	About a million eyed eggs, planted along lake shores by Klawak hatchery.
1913	sockeye	1,970,000	Uganik Lake (Afognak Is.)	Eggs eyed in troughs on site. Field station, Afognak hatchery. Planted in lagoons of Uganik Lake.
1913	sockeye	2,180,000	Eagle Lake (Kodiak Is.)	Eggs eyed in troughs on site. Field station Afognak hatchery. Planted in Eagle Lake.
1916	pink	325,000	Ketchikan Creek	Green eggs planted in creek. Field station of Yes Bay hatchery.
1916	pink	29,000	Smeaton Bay	Green eggs planted in nearby spawning beds. Field station, Yes Bay hatchery.
1917	sockeye	2,712,000	Seal Bay (Afognak Is.)	Eyed eggs planted in Little Seal Bay Creek. Field station of Afognak hatchery.
1917	pink	...	Seal Bay	Unknown part of 2,560,000 eggs planted in lake and remainder sent to Afognak hatchery.
1920	sockeye	350,000	Litnik Lake (Afognak)	Planted eyed in 3 lakes in Yes Bay, by Yes Bay hatchery.

Table 2 — Continued

Year	Species	Number of Eggs	Source	Disposition
1921	sockeye	150,000	McDonald Lake (Yes Bay)	Planted eyed in 2 lakes in Yes Bay by Yes Bay hatchery.
1928	pink	50,000	Eva Lake	Green eggs planted by A.J. Sprague, Territorial Fish Commission, in barren lake in Portage Bay.
1928	pink	70,000	Eva Lake	Green eggs planted by Sprague in Cascade Bay Lakes.
1928	pink	50,000	Eva Lake	Green eggs planted by Sprague in Saginaw Bay tributaries.
1928	pink	50,000	Eva Lake	Green eggs planted by Sprague in Tebenkof Bay.
1928	pink	40,000	Eva Lake	Green eggs planted by Sprague in Hamilton River.
1928	pink	10,000	Eva Lake	Green eggs planted by Sprague in Big Johns Harbor.
1928	pink	30,000	Eva Lake	Green eggs planted by Sprague above falls in Hamilton Bay.
1953	sockeye	20 females	Naha River (Heckman Lake)	Green eggs planted by C.L. Anderson, Alaska Department of Fisheries, in Old Frank's Lake system.
1953	sockeye	23 females	Walker Creek (McDonald Lake)	Green eggs planted by Anderson in Old Frank's Lake system.
1953	coho	13 females	Reflection Lake	Green eggs planted by Anderson in inlet stream of small lake, adjacent to Wolf Creek.

The actual procedures used by Sprague to plant the eyed eggs is not known. When the Territory again began planting eyed eggs during the 1950s as part of the program to introduce salmon to barren systems, the procedures were recorded and are described in that chapter.

Barren lake stocking was first introduced by John C. Callbreath in 1895-1896 when he planted sockeye sac-fry into Burnett and Navy lakes in Southeast Alaska. He selected both systems because there were barrier falls which kept anadromous fish from entering. Callbreath's diaries do not cover the years he might have checked for returns of adult spawners.

Another barren lake transplant was made in 1920-1921 by Superintendent C.H. Van Atta at the Yes Bay hatchery. During December, 1920, 350,000 eyed eggs, within about five days from hatching, were planted in the gravel of three lakes near Yes Bay — Cannery Lake, Round Lake and Lake #2, none of which are identifiable today. These bodies of water did not support a run of salmon, but were considered by Van Atta as suited for the purpose. It was believed that such bodies of water had sufficient food, and the fish would escape to the sea without injury during periods of high water.[56]

The lakes were visited in mid-May, 1921, and it was written in the log: "We saw fry in the lakes but could not be certain they were sockeye as there are numerous sticklebacks in the lake and all fry seem so shy that we could not get close enough to make out what they were."[57] It was reported in published reports that there were several schools of 25 to 75 sockeye fingerlings from 1 to 1.5 inches long.[58]

The following year (1922) 150,000 eyed sockeye eggs were planted in Round Lake and Lake #2, but during a visit the following May, the water was too high and too dark to see any fish. In June no sockeye fry were seen, but many cutthroat were jumping in the lake.

A lake stocking program was carried on by Sprague from 1917 to 1920. He planted eyed pink salmon eggs in a barren stream and tributary to Chilkoot Lake and reported a return. His biggest project under the auspices of the Territorial Fish Commission was planting of eyed coho and pink eggs and coho fry in Baranof Lake above Warm Springs Bay on Baranof Island. He reported a return of 40,000 adults at the base of the falls.[59]

Introduction of salmon into barren lake systems became the major thrust of a Territorial program in the 1950s, and that program concentrated on developing runs to barren lakes and construction of fishways or removal of the barriers to open the lakes to future generations of returning salmon. This end was pursued at Afognak Island in the Pauls Lake system and at Fraser Lake on Kodiak Island. It was utilized at Bakewell Lake south of Ketchikan and was attempted in several other Southeast Alaska lakes.

The Kitoi Bay Research Station was started in the mid-1950s because the Territorial Department of Fisheries felt that scientific knowledge was lacking on how to, how much to, what to, and where to plant in barren lakes. Most of these lake stocking programs and the research of the Kitoi Bay Research Station continued after statehood.

The third experimental field was started by John Pease Babcock, assistant to the Commissioner of Fisheries of the Province of British Columbia in 1910. He advanced what was considered a novel idea: that freshly fertilized eggs buried immediately under sand and gravel would produce strong healthy fry at less cost than under existing hatchery methods. Fry so produced, he stated, were stronger and more capable of resisting the attacks of their active enemies.[60]

F.M. Chamberlain, with the Bureau of Fisheries, tested Babcock's experiments in 1912 at the Yes Bay hatchery. Two sets of conditions were used. The first eggs were fertilized and washed to avoid burying matter subject to decomposition. They were distributed in egg baskets, left a few minutes and then small sand and gravel (matter which had been screened through basket wire) were scattered over them. This tended to displace the eggs and it was necessary to introduce small boulders so crevices would be formed in which eggs could lodge. The basket was then buried in the boulders just below the rack in the hatchery stream, a place known to remain open all winter. It was set down with the rim level with the stream bed and filled with successive layers of coarse and fine gravel.

In the second experiment, a Williamson trough was used. In the trough, a layer of gravel was put on the bottom, and a dyke of large boulders built a foot from the lower end leaving a space into which fry might drift after hatching. Free percolation of water was obtained. The uppermost eggs were covered by at least an inch of gravel and sand.

The basket in the stream was opened on March 17, 1913, and all eggs were dead, none having reached the hatching stage. Eggs seemed to be smothered by the fine silt washed down during high water.[61] The experimental eggs placed in the hatchery trough were examined on the same day. Commencing at the lower end of the trough, the rocks and gravel were carefully removed to ascertain the depth to which eggs and fish could be found. When the rocks were comparatively larger and loose, and the eggs were not put in a large mass, 90 percent hatched and were alive to a depth of two inches. When eggs were packed in fine sand, 75 percent were found dead at a depth over one inch. When the rocks and gravel were so arranged that the water could seep through, live fish and eggs were found at a depth of three to four inches. Taking the average of the trough, about 45 percent of the eggs hatched out.

It was observed that in places a larger number of fry could be found hatched out, but dead, which would seem to indicate there was sufficient aerated water to incubate the eggs, but not enough to sustain the young fish. At the end of

April the fry produced in gravel appeared to be doing well. The experiment was not repeated, and the subsequent development of the technique for producing fry from gravel incubation was largely confined to British Columbia. At the 1919 meeting of the American Fisheries Society, Alexander Robertson read a paper entitled "Hatching Fry in Gravel" which described detailed procedures for successful incubation.[62]

Edwin Wentworth, superintendent of the Afognak Station, picked up on the idea and ran experiments in 1919-1920. The experiment was described as follows:

> "Eggs to the number of 5,500,000 were placed in two old hatching troughs, mixed thoroughly with gravel, and bedded in the bottom of the creek. The troughs were fitted with a cover and water was introduced through a 1-inch pipe leading from a small dam, so that there would be a gradual flow through each trough. The intake of this pipe was carefully protected by a screen.
>
> "On opening the troughs in the spring it was found that very few of the eggs had hatched. Though the winter was a mild one it was probable the eggs had been frozen, as there had been a very light fall of snow, leaving the creek unprotected, and frost may have penetrated to a greater depth than in a season of lower temperatures. Eggs placed in gravel boxes patterned after the Robertson plan and held in the hatchery developed nicely, and the majority of the fry remained in the gravel until the yolk sac was absorbed."[63]

The entire lot was lost the following year. When the hatching trough was opened, it was found that development had been arrested at the eyed stage, and it was apparent that the water circulation had not been adequate. This was the last experiment with incubating salmon eggs in gravel recorded prior to statehood.

TRANSFER OF EGGS

It was the policy of the Bureau of Fisheries to allow transfer of salmon eggs sanctioned by them from one state or Territory to another. Most of the transfer of salmon eggs in Alaska was out of the Territory. The exceptions were the egg shipments to Alaska of eastern brook trout, rainbow trout and chinook and coho salmon. All chinook egg transfers were in attempts to establish a fall run of that species.

The majority of the transfer of salmon eggs from Alaska was done by the Bureau from their two hatcheries at Yes Bay and Afognak. Tables 3 and 4 show the transfers of eyed salmon eggs from each hatchery. These are incomplete but are compiled from available records. Kemmerich remembers shipments of pink eggs in the fall of 1923 from Yes Bay hatchery.[64] Harry Baer, attached to the

Table 3. Eyed Egg Transfers from Afognak Hatchery, 1914-1932

Year	Number of Eggs	Species	State	Hatchery
1914[a]	5,500,000	pink	Washington	...
	7,000,000	pink	Maine	...
1916[b]	4,096,000	pink	Maine	Craigs Brook
	4,106,752	pink	Washington	Green Lake
	4,000,000	pink	Washington	Birdsview
	2,000	pink	Washington	Duckabush
	2,000,000	pink	Washington	Quilcene
1917	10,000,000	sockeye	Canada	Harrison Lake
	5,000,000	sockeye	Washington	Quinault (Grays Harbor)
	3,000,000	sockeye	Oregon	Bonneville
1918	20,700,000	sockeye	Canada	Harrison Lake
	5,000,000	sockeye	Washington	Quinault
1919	7,000,000	sockeye	Washington	Quinault
	3,000,000	sockeye	Oregon	Bonneville
	3,300,000	sockeye	Alaska	Ikatan Lake tributaries
1920	2,340,500	sockeye	Washington	Quinault
	3,000,000	sockeye	Oregon	Bonneville
	5,000,000	sockeye	Alaska	Yes Bay
1921	5,200,000	sockeye	Oregon	Bonneville
1922	5,098,936	sockeye	Alaska	Auke Lake/Mendenhall Bar
	5,045,000	sockeye	Oregon	Bonneville
	534,464	sockeye	Washington	...
	278,616	sockeye	Washington	Birdsview
1925	1,023,360	steelhead	Washington	Puget Sound area
1926	3,402,000	sockeye	Washington	...
	3,617,000	pink	Washington	Quilcene
	790,400	steelhead	Washington	...
1928	1,851,550	sockeye	Washington	...
	2,300,800	pink	Washington	...
1929	4,553,200	sockeye	Washington	...
1930	10,155,776	pink	Washington	...
1931[c]				
	2,731,424	Dolly Varden	Washington	...
1932	2,228,000	pink	Washington	...
	154,000	Dolly Varden	Washington	...

[a] from Litnik River, Karluk and Uganik Bay

[b] from Litnik River, Seal Bay and Uganik Bay

[c] from Buskin River

Table 4. Eyed Egg Transfers from Yes Bay Hatchery, 1910-1932

Year	Number of Eggs	Species	State	Hatchery
1910	100,000	pink	Washington	Baker Lake
	1,500,000	sockeye	Oregon	Bonneville
1911	2,000,000	sockeye	Oregon	Bonneville
1912	2,000,000	sockeye	Oregon	Bonneville
1913	2,000,000	sockeye	Oregon	Bonneville
1914	3,000,000	sockeye	Oregon	Bonneville
	2,000,000	pink[a]	Washington	Birdsview
1915	3,000,000	sockeye	Oregon	...
	50,000	sockeye	Washington	Quinault Lake
	50,000	sockeye	Canada	Anderson Lake
	15,000,000	sockeye	Alaska	Afognak
1916	2,000,000	sockeye	Oregon	Bonneville
1918	3,440,100	sockeye	Oregon	Bonneville
	1,059,000	sockeye	Oregon	Little White Salmon
	406,000	pink	Washington	Birdsview
1925	5,645,000	sockeye	Washington	Baker Lake
	300,760	sockeye	Oregon	...
1926	5,241,130	sockeye	Washington	Baker Lake
	1,717,000	sockeye	Alaska	Ketchikan Territorial
1928	2,493,000	sockeye	Washington	...
	2,038,000	pink	Washington	...
1929	1,021,000	pink	Washington	Auburn
1930	6,107,000	pink	Washington	Auburn
	1,187,000	sockeye	Washington	Birdsview
	1,868,000	sockeye	Washington	...
1931	3,144,960	sockeye	Oregon	...
1932	379,904	pink	Washington	...
	3,010,650	sockeye	Oregon	...

[a] from eggs taken at Mink Arm, Boca de Quadra

Afognak hatchery, spent each spring between 1929 and 1932 collecting steelhead eggs in the Karluk River.[65] The eggs were eyed in hatchery troughs on site (halfway between the mouth of the river and the lake), packed, and delivered to the patrol boat at the head of Larsen Bay. He was never told the disposition. How many other unrecorded shipments were prepared at the two stations will probably never be known.

The majority of the early transfers involved specific experimental programs. The earliest were in conjunction with the Bonneville, Oregon, hatchery operated by the Oregon Department of Fisheries. The first attempt to restock the Columbia River with sockeye salmon was made in 1910, and 1.5 million eyed eggs from Yes Bay were sent to the Bonneville hatchery. Other shipments of both pink and sockeye salmon eggs were made to Washington and Oregon. Queries to the Oregon Fish and Game Commission and the Washington State hatchery system did not locate records on receipt of Alaskan eggs.

Alaskan salmon were used in the continued efforts by the Bureau of Fisheries and various state fish commissions to introduce Pacific Coast salmon in eastern waters. The first pink eggs were transferred in 1913 from several Pacific Coast hatcheries to the Bureau's Craigs Brook, Maine, hatchery. Shipments were made for successive years and the 1914 and 1916 shipments included Alaskan eggs.[66] The Alaskan shipments were from several pink salmon egg sources: in 1914 from Uganik Bay on Afognak Island, Karluk River on Kodiak Island, and from the home stream, Litnik River, at the Afognak hatchery; and in 1916 from Uganik Bay, Seal Bay (Afognak Island) and Litnik River.

The other major program to rehabilitate depleted runs, was on the Fraser River in British Columbia in 1917 and 1918. At that time the salmon runs to the river were considered seriously exhausted. A series of hatcheries had been started after the Hells Gate Canyon blockage of 1913, but the runs were so diminished that some hatcheries were closed due to lack of eggs. After correspondence between Canada and the United States, a shipment of Afognak Island sockeye eggs was made to a British Columbia hatchery.[67] The initial shipment by ocean steamer in 1917 was poorly packed and handled enroute and nearly all eggs were lost. Another shipment was sent in 1918, which arrived in good condition.

The other Canadian shipment was made in 1915 for research purposes. The intention was to mark resultant sockeye fingerlings to prove that adult fish would return to the streams in which they had passed their early existence, no matter where the eggs were taken. As it turned out, the young fish were not strong enough to survive marking and were liberated without marking.[68]

Only once did the Bureau reluctantly allow another agency to take Alaskan eggs. The Washington State Fish Commission, in 1922, requested permission to take pink salmon eggs in Southeast and Central Alaska. The request stated that approximately half of the eggs, once eyed, would be returned to the streams from which they were secured. The Bureau recommended that permission be denied because the fisheries of the Territory were considered to be in a precarious state and the export of eggs would add to the already existing danger of serious depletion of the streams.

The Bureau's representative in charge of the Southeast district telegraphed Washington, D.C., officials in mid-August, that a barricade had been placed in Anan Creek and preparations to take eggs were being made with apparently no plan to return eyed eggs to the stream. Then the Bureau's agent in Cordova reported that over ten million pink eggs had been collected by representatives of the Washington State Fish Commission at Irish Cove, on Fidalgo Bay, Prince William Sound. He felt these eggs probably would be a total loss because water from the vicinity of an abandoned copper mine was being used. Instructions were issued that all eggs on hand be planted and all operations cease. In the meantime, the Commission had moved to Whalen Bay, where the water proved to be more suitable and further collections were made.

Finally, an agreement was reached with the Washington State Fish Commission that the operation at Anan Creek would cease entirely; that the healthy eggs already secured on Prince William Sound would be eyed for shipment and no further collections made, and that the expedition would withdraw entirely from Alaska. Final reports show that 14,571,708 eyed pink eggs were shipped. Representatives of the Washington State Fish Commission collected pink eggs again in 1924 and maintained an eyeing station at the head of Fidalgo Bay on Prince William Sound.

All shipments were made from Alaska as eyed eggs. When Alaskan shipments were first started, the techniques for shipping eggs were well developed. Packing boxes had been designed and described.[69] Basically the box was two feet square and one foot deep. Trays of eggs were placed between a layer of mosquito netting and a thick layer of moss and stacked to the middle of the box. Here a firm partition was fastened in, and then packing resumed. A lid was laid on top. Two of these boxes were placed in a wooden crate made large enough for a space of three inches on all sides of the boxes. This space was filled with hay (or moss in Alaska) to protect the eggs against changes of temperature. In the middle of the crate an open space about four inches in depth was left between the two boxes of eggs for ice. The shipment of eggs in 1917 to British Columbia was lost because there was no ice to keep in the crates.

By 1920, O'Malley, in his manual, advocated the Atkins-Dinsmore shipping case, which he describes in detail.[70] It eliminated the use of ice, but the packing had to be done in temperatures below 50 °F and all packing materials kept in a place slightly below the freezing point until ready for use. This method was best for shipping eggs in cold weather, but in warm weather an ice hopper could be added, which dripped water down through the eggs.

PREDATOR CONTROL

Predator control was considered necessary by early hatchery operators. It was the common belief that Dolly Varden, sometimes called the salmon trout or char, caused great destruction to the salmon population by devouring the eggs in the fall and the young fry in the spring, although prior to 1906 these fish were protected by law. Sculpins were also considered predatory. It was believed by some that yearling coho were predators of sockeye fry. Adult coho were prevented from returning to some systems such as Heckman Lake and the Fortmann hatchery.

In Alaska it became the practice for hatchery personnel to capture and destroy fish considered predators prior to release of fry in the lake. This was done by several means, but the most popular was by use of a seine or gillnet. Another method was used at Fortmann hatchery in 1907. A baited wire net trap was used to capture 40,000 sculpin during the year — 2,700 in one night. It also took young trout and char. At Yes Bay hatchery, during the spawn take, the Dolly Varden would collect around the outside of the seine. These were captured and destroyed. Another common practice was to use dynamite in an area before fry were released.

A dam or rack to prevent upstream migration of predators was often used. The Territorial Fish Commission constructed these on outlets of many lakes and used them to count adult salmon. Callbreath was adamant in his belief that predators must be kept from the spawning and rearing grounds and utilized a dam at which he passed only desired species.

At most places where Dolly Varden were captured, they were left on the bank to die and the carcasses disposed of in the lake. The federal hatcheries utilized these predators: at Afognak they were salted for fish food; at the Yes Bay hatchery they were ground and used as fresh food for the fry.

The vast numbers of Dolly Varden which were destroyed through these programs each year are not accurately recorded. Isolated accounts give some idea of the numbers captured and destroyed. At Afognak, from 1924 to 1928, a period of four years, a total of 117,050 Dolly Varden were destroyed. In three years 40,644 Dolly Varden were prevented from entering Eyak Lake by the weir operated at the outlet.

Some hatchery operators considered ducks and gulls as predators also, and these were systematically destroyed during spawning season and at the time fry migrated from the lake.

EVALUATION

Evaluation of the success of hatchery programs, as defined today, was not attempted by the federal and private hatcheries in Alaska. Only the Territorial programs, both in the 1920s and 1950s, tried to authenticate the returns as originating from hatchery efforts.

At the federal hatcheries an occasional attempt was made to determine the number of yearlings leaving the lakes in the spring. These attempts, usually unsuccessful, have been described in detail in the station histories of the Yes Bay and Afognak stations. Only one report of an outmigrant study at a private hatchery was located. This was Fortmann, but the study was done by a Bureau of Fisheries employee, F.M. Chamberlain, in 1903-1904, and is discussed in the Fortmann station history.

In Alaska the first fin marking of salmon fry and fingerlings was not initiated to prove the efficiency of the hatchery program. It was to learn the salmon's life cycle and prove the parent stream theory which generated the desire to apply distinctive marks to young fish. The first hatchery operator to record marking of sockeye in Alaska was James A. Richardson, superintendent at Karluk.

Richardson believed in the parent region theory — that fish returned to a region, not a specific stream. To prove he was correct, Richardson, in 1897, cut one fin from 513 fingerlings. The fish were from eggs taken in 1896. The fry had been held in rearing ponds until they were from 1-¾ to 4-½ inches in length. He held a portion of the marked fish for two months longer and reported, "They do not seem to be the worse for the loss of a fin."[71] He released the fish in the fall of 1897.

Experiments continued at Karluk where some fry were marked either by removal of the adipose fin or by notching one of the other fins. Mature salmon with these markings were taken in the Karluk fisheries and Richardson reported in 1901 that several of the fish he had marked returned.[72] The information conveyed was so meager and the record of marks so obscured by a repetition of the same mark during different years, that even in 1905 Alaska Packers Association officials, who owned the hatchery, knew it was impossible to form any proper conclusions.[73]

Another fin clipping program which was designed to prove the parent stream theory was undertaken at Klawak in 1899. Here the cannery superintendent, H.F. Swift (who also supervised the hatchery), wanted to prove that sockeye came back to a single stream, in this case the Klawak River. He used 12,000 yearlings which were about 18 months old and ready to go to sea, and cut the adipose fin smooth to the back with scissors. In 1902, Swift reported to Coyle, the assistant fisheries agent, that 150 of these fish were found in the Klawock cannery catch without any special search when the fish were brought to the cannery.[74] The fish averaged 8 pounds and 32 inches in length, and the mark of the fin was grown over smooth to the skin. Other sockeye salmon streams in the area were checked and only one marked fish was found outside the Klawak River.

Swift reported that three such experiments were done with a period of one or two years intervening between markings. One fisheries agent wrote that the yearlings used for marking "must have been in part, presumably much the smaller part, the product of natural spawning."[75] The number of fish used in 1899 varies according to the source from 12,000 to 1,000. The other two experiments used about 2,000 fish. It was reported "a return of about 20 percent is claimed in each of the first two cases, and about 5 percent in the third. The return in every case occurring the third year after marking."[76]

The other markings on salmon fingerlings were conducted by Bureau of Fisheries employees at private hatcheries or at the federal hatcheries.[77] The major, best documented, but confusing, study was done by Chamberlain.[78] His salmon marking experiment had two purposes. One was to determine the exact age of maturity of sockeye salmon, and the second was to obtain results which showed that the mutilated parts would not regenerate, something which had not been proven at that time.

When Chamberlain first began his studies in 1903, only a few attempts at marking Pacific salmon had been made by the Bureau of Fisheries. One of these was that of Waldo Hubbard on the Clackamas River. He removed the adipose fin from chinook salmon to determine their maturity. In Chamberlain's first experiment in August, 1903, he used 1,600 sockeye fry, reared for that purpose at the Fortmann hatchery from eggs taken in 1902. These were marked by excising both ventral fins with fine curved scissors. It was the practice to mark two fins so that the probability of overlooking the mutilations in the adult would be minimized, and at the same time the chance of mistaking accidental marks would be lessened. The fry were released as soon as marked at a point in the Naha River just above Heckman Lake. At that time they were about three months old.

No adult sockeye lacking ventral fins were taken at the Fortmann hatchery until fall of 1906 and then only two fish. From 50 to 100 such fin marked fish were reported by the superintendent of the Yes Bay hatchery in the spawning run of 1906, some 30 miles from the Naha system. No such marked fish were seen in 1907 at Yes Bay hatchery, but 12 to 15 were found with one ventral removed. At Fortmann hatchery the 1907 spawning run yielded 13 sockeye with both ventrals gone and one with a single ventral. A return of five at Fortmann and three at Yes Bay with both ventrals missing occurred in 1908.

A summary published in the 1910 annual Alaska fisheries report showed returns found for all years:[79]

YEARS	Naha	Yes Bay	Karluk	Total	Age of Fish
1906	2	...	...	2	3
1907	13	...	...	13	4
1908	5	3	...	8	5
1909	...	4	1	5	6
1910	1	10	1	12	7
TOTAL	21	17	2	40	...

One of the ten fish credited to Yes Bay hatchery in 1910 was caught in the bay, all others consigned to Yes Bay were taken at the hatchery. The numbers shown do not include fish returning with only one ventral fin, nor the 50 to 100 fish which supposedly returned to Yes Bay in 1906, none being preserved for specimens. The Karluk return was included because, it must be remembered, it was still not known for sure if sockeye salmon returned to their parent stream.

Various analyses of this marking experiment appeared in the annual Alaska fisheries reports. In 1907, after only two years returns, and without the confusion of the Karluk markings, this analysis was made:

> "... the return is nearly one percent on fry liberated at the age of about 3 months. Most of these returned in 1907, and indicate 4-½ years as the approximate age of the redfish from the time of hatching to sexual maturity. That all individuals of a given hatching do not mature in the same year is not improbable and is indicated by the nature of the Naha marked fish at Yes Bay in 1906. The same evidence also indicates that while a part of a given hatch of salmon may return to the parent stream, a greater part may go to other streams of the region."[80]

This analysis was made in 1908:

> "The return has covered three successive seasons, 1906 to 1908 inclusive, and amounts now, counting only certainly marked specimens to 23 salmon or 1.4 percent of the number marked and liberated. The importance of returns from these markings lie in adding certainty to the identification of marked fish with those returning, and in the spreading of the return over three successive years and into other waters than the home stream, rather than in any demonstration of the numerical proportion of fish returning and the consequent effect of similar plants on the general salmon supply. Considering the indeterminate numbers of incompletely identified salmon taken at Yes Bay in 1906 and supposed to bear the

mark, and the returns which have undoubtedly occurred unobserved, it is well to not make much inference concerning the size of the total run, save that the percentage given is certainly a minimum."[81]

After 1908 the experiment becomes clouded because several fish with both ventrals missing were discovered at Karluk. One of these was forwarded to the Bureau of Fisheries and, upon examination, was identified with fish so marked by Chamberlain in Southeast Alaska. This was before the time Gilbert began to read fish scales, so it is unknown how this determination was made.

None of the Karluk reports to the Bureau of Fisheries mention marking experiments at Karluk after Richardson left Karluk. Yet a hatchery employee remembered that in 1909 "Many were marked by clipping one or more of their fins, so they could be identified when they returned four years later."[82] If this practice was continued up until at least 1909, this could explain the marked fish at Karluk.

The inclusion of the Karluk marks and similar marks found as late as 1911, caused great confusion in the scientific world. Gilbert, in 1914, questioned the fact that eggs taken in 1902, (with marked fish released in 1903):

"... had returns in 1906 and 1907 and continued to be reported each year since that date including the summer of 1911. If these results are accepted, individuals may even reach the age of nine years before maturing; but this is a conclusion so in variance with other evidence and with general probability that complete collaboration is required."[83]

The Bureau of Fisheries was greatly concerned that marked fish other than those in the Fortmann hatchery experiment were found. Cobb, in his 1910 visit to the Quadra hatchery, reported that seven marked sockeye had been taken at the hatchery: two with anal fin removed and five with a section of the middle of the anal fin cut out.[84] The hatchery at Quadra had been inoperative from 1904-1907 and there was no record of fins clipped in 1908.

At Yes Bay hatchery, in 1907, in addition to those fish previously reported with one ventral clip, eight to ten had an adipose clip, and three to five a pectoral clip. In 1908, two adipose and one pectoral clip were located in addition to the ventral clips. In 1909 similar numbers were found.

At Karta Bay, on eastern Prince of Wales Island in 1909, an Indian caught 17 sockeyes from which the adipose fin was clipped. He reported this to the superintendent of the cannery at Loring, his catch was kept separate, and five more were discovered.

All of these marks would have been applied before it was decided that marking experiments must be conducted or monitored by the Bureau. Under date of March 17, 1908, the Secretary of Commerce and Labor sent to owners and operators of salmon hatcheries in Alaska a notice which said, in part:

> "The department. . . intends to conduct through a term of years a series of investigations of the fisheries, particularly by marking and liberating salmon fry artificially hatched, and examining the return of adults. Such markings have hitherto been made by various people in Alaska. It is necessary in order that confusion of results may be avoided that no further experiments of this sort be undertaken by the commercial hatcheries, or that none be made save by the permission of this department. . . ."[85]

In 1912 the Fortmann and Yes Bay hatcheries found what was considered to be a remarkable number of marked salmon. Chamberlain reported:

> "From the former station [Yes Bay] the superintendent reports having taken 28 females and 13 males with both ventrals missing and 6 females and 4 males with one ventral gone. He adds that he believes had the examination of the males in the course of spawning been as thorough as in the females, there would have been as many males as females noted. About a dozen of these marked fish were reported from the Loring [Fortmann] hatchery up to the end of October."[86]

Chamberlain examined the scales of three of the fish and found that all appeared to be four year fish. "There is always a shade of doubt regarding the scale rings of spawning fish but in no case can these fish be adjudged to be more than 5 years old and in all probability only 4 years."[87]

He also observed:

> "To attribute these fish to the marking done in 1903 is absurd. Aside from the inherent improbability of a second group appearing in this way at such a distant interval from the first return in 1906, the record of the scales is final evidence of the real age of the fish."[88]

Chamberlain had several theories to account for the presence of the marked fish. One was that the disappearance of the fins was due to a natural cause and the second, the fish were marked by a human agency. Against the first proposition was that no fins except the ventrals were damaged and none of the damage involved adjacent structures. His detailed arguments can be found in the 1912 annual Alaskan fisheries report.[89]

This left the supposition that the fins had been clipped when the fish were young. The superintendents of both federal hatcheries asserted they had not marked the fish, and insisted that it would be impossible to mark any number of fry at the establishments without their knowledge. Yet marking was done in 1909 at Yes Bay hatchery. The logbook for that year notes that on November 27 three lots each of 60 fingerlings were marked. None of the marking was the same as the fish returned. All included removal of the adipose fin, in conjunction with a dorsal, a half dorsal and a ventral clip. It was unusual for the logbook to include mention of such a project and no follow-up was recorded.

Chamberlain then speculated on the Quadra and Hetta hatcheries, both of which normally ceased operations immediately after the last eggs were hatched and sac-fry released. They would not have had the necessary number of fingerlings. Chamberlain also felt that anyone else would have a difficult time obtaining sockeye fry in the numbers necessary except in a hatchery situation.

Chamberlain reached no conclusion as to the probable origin of the marked fish. The only motive he could discern for the marking was an attempt by a private party to perpetuate a scientific experiment started and ceased by the Bureau of Fisheries in Alaska. He also thought someone might have wanted to show a larger return from the number officially marked and announced, or to make a false showing as to the efficiency of artificial propagation by fostering a belief that the returns were delayed over a long series of years.[90]

But the confused conclusions of Chamberlain's study at Naha River did not dampen his curiosity concerning the potential of marking salmon. He learned several things from his 1903 experience upon which he felt further study should be done. He had marked fry in his experiment and in 1904 he attempted to mark yearlings, several thousand of which he trapped as they migrated into Naha Bay. No record was ever made of one of these fish returning. He believed the handling resulted in a great loss and the percentage of survival was too small to obtain notice.[91]

Chamberlain speculated that, if the fish could be marked when they were under two inches, they would be less susceptible to injury. "The experiment of anaesthetizing has not been made," he said in 1907.[92] He turned his attention to this in 1912. He tried using chloroform in an experiment at the Yes Bay hatchery. " The first test of the anaesthetic resulted in the hasty death of about a dozen fish. The anaesthetizing was a success, but the resuscitation was not."[93] In subsequent tests he found that a teaspoon of chloroform in a bucket of water, about three gallons, seemed a usable solution. The fish showed uneasiness after two or three minutes, and in four or five could be taken in hand without resistance. All fish survived without apparent harm.

The main reason Chamberlain was interested in controlling the struggle of the small fish, was to test another theory he had on inserting what was the forerunner of today's coded wire tag. A piece of silver wire was flattened on an anvil to make a ribbon about two millimeters in breadth and was cut into sections approximately one centimeter in length. The corners were rounded by a file. He wrote "The silver pieces may be sufficiently large to admit of their having stamped with small figures and thus, afford a means of positively identifying individuals . . . The excision of the fins gives an external mark that will differentiate these fish from the mass."[94] Essentially this is a description of a coded wire tag, except stainless is used, and is one millimeter in length. It is inserted into the cartilage of the head, and the adipose fin is clipped.

Chamberlain's silver marker was inserted into the fleshy part of the body forward of the dorsal fin in both sockeye and coho fingerlings, varying from two to five or six inches in length. The fish were held to determine the effects of the tagging and fin clipping. Chamberlain found that of 35 fry examined, 9 carried the tag. One was completely healed, five had healed but a sharp corner of the metal had cut or irritated the skin at another place, then had healed but later developed infections which would probably lead to death. An uncertain number, four or five at least, had lost the tag and in about 12 it was impossible to tell without killing the fish, if they had been tagged. Chamberlain concluded ". . . but considering the conditions under which it [the experiment] was made, the indications are that the method is worthy of a better test."[95] This technique was not perfected until the late 1960s.

Chamberlain's 1912 experiment was the last recorded effort by the Bureau in Alaska to mark fish. The most extensive marking programs in Alaska prior to statehood were carried on by the Territory in the 1920s and the 1950s. Both were designed to substantiate returns of hatchery fish used in development of fish culture techniques. The 1920s program is covered in detail in the history of the Ketchikan hatchery. The methods used are not recorded except that fins were clipped.[96] This program, under Edwin Wentworth, was the first to use different clips each year so that fish from each release would be identified. The 1950s program, especially at the Deer Mountain hatchery in Ketchikan, also involved fin clips, but again procedures were not recorded.

The 1920s program ceased before analysis could be made and fish captured in the fisheries. If fish returned to streams of release, there was no one there to record it. The continuity of the 1950s program was broken by the change from Territorial to State controlled fisheries and many of those records have become misplaced in the transfer.

To better understand how the fish culture procedures discussed in this chapter were used at the various Alaskan hatcheries, histories of each hatchery have been developed.

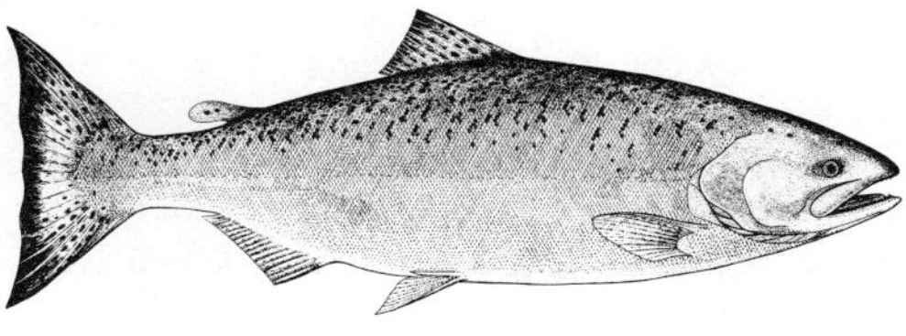

Notes

1. F.M. Chamberlain, "Habits of Salmon," *Pacific Fisherman,* Vol. II, No. 1, January 1904.

2. F.M. Chamberlain, *Some Observations on Salmon and Trout in Alaska,* Bureau of Fisheries Document No. 627 (Washington: Government Printing Office, 1907), p. 31.

3. The study of scales was not a new idea as it had been experimentally tested and approved by the Fisheries Board of Scotland in the case of Atlantic salmon. It has been shown to be applicable to various species of trout.

4. Chamberlain, *Some Observations on Salmon and Trout in Alaska,* p. 25.

5. Charles H. Gilbert, *Contributions to the Life History of the Sockeye Salmon,* Paper No. 3 (British Columbia Fisheries Department, 1915), p. 527.

6. John J. Brice, *A Manual of Fish-Culture Based on the Methods of the United States Commission of Fish and Fisheries* (Washington: Government Printing Office, 1898).

7. Ibid, p. 6.

8. U.S. Commission of Fish and Fisheries, *A Manual of Fish Culture Based on the Methods of the United States Commission of Fish and Fisheries,* Revised Edition (Washington: Government Printing Office, 1900).

9. Ward T. Bower, "Fish Culture in Alaska," *Alaska Fisheries and Fur Industries in 1911,* Bureau of Fisheries Document No. 766 (Washington: Government Printing Office, 1912).

10. Ibid, p. 80.

11. Henry O'Malley, *Artificial Propagation of the Salmon of the Pacific Coast,* Bureau of Fisheries Document No. 979 (Washington: Government Printing Office, 1920).

12. John N. Cobb, *Pacific Salmon Fisheries,* Bureau of Fisheries Document No. 839 (Washington: Government Printing Office, 1917)
John N. Cobb, *Pacific Salmon Fisheries,* Bureau of Fisheries Document No. 1092 (Washington: Government Printing Office, 1921).

13. A.B. Alexander, "Notes of A.B. Alexander, Fishery Expert," Records concerning the Steamer *Albatross,* General Records, Records of the United States Fish Commission and Bureau of Fisheries, RG 22, NA.

14. U.S. Congress, Senate, Subcommittee of the Committee on Fisheries, *S. 5856, A Bill to Amend An Act for the Protection and Regulation of the Fisheries of Alaska,* Hearings, 62nd Congress, 2nd Session, p. 219.

15. J.N. Wisner, "Alaska Report, 1903," File "Hatchery Inspection Report, 1903-1910," General Records, Division of Alaska Fisheries, RG 22, NA.

16. Ibid.

17. O'Malley, p. 28.

18. A more detailed explanation with diagrams of Williamson troughs and baskets can be found in O'Malley, pp. 20-22.

19. J.T. Bowen, "A History of Fish Culture as Related to the Development of Fishery Programs," *A Centruy of Fisheries in North America,* Norman G. Benson, editor, Special Publication No. 7, American Fisheries Society (Washington: American Fisheries Society, 1970), p. 76

20. John W. Titcomb, "Experiments in Fish Culture during the Past Year in the Bureau of Fisheries," *Transactions of the American Fisheries Society, 1905,* Vol. XXXIV (Appleton, Wisconsin: Post Publishing Co., 1906), pp. 69-71.

21. Brice, p. 18

22. O'Malley, p. 20

23. Ibid, p. 22

24. Brice, p. 16.

25. Titcomb, p. 60

26. Bower, p. 80

27. O'Malley, p. 19

28. Statement of Mark Meyer, former superintendent, Quadra hatchery, in personal interview, Washougal, Washington, July 31, 1979.

29. Brice, pp. 16-17.

30. Ibid.

31. O'Malley, p. 20.

32. Bower, p. 81.

33. Ibid, p. 82.

34. Claudius Wallich, "A Method of Recording Egg Development for Use of Fish Culturists," appendixes of the *Report of the Commissioner of Fish and Fisheries for Year Ending June 30, 1900* (Washington: Government Printing Office, 1901), pp. 185-195.

35. O'Malley, p. 23.

36. Titcomb, p. 58.

37. Henry O'Malley, "Salt Solution as an Aid to Fish Culture," *Transactions of the American Fisheries Society, 1905,* Vol. XXXIV (Appleton, Wisconsin: Post Publishing Company, 1906), pp. 49-51.

38. Bower, pp. 84-85.

39. O'Malley, *Artificial Propagation of the Salmon of the Pacific Coast,* p. 23.

40. Bower, p. 87.

41. Ibid.

42. Glen C. Leach, *Propagation and Distribution of Food Fishes, 1921,* Bureau of Fisheries Document No. 912 (Washington: Government Printing Office, 1922), pp. 30-31.

43. Statement of Alphonse Kemmerich, former superintendent Yes Bay hatchery, in personal interview, Woodburn, Oregon, July 30, 1979.

44. Leach, *Propagation and Distribution of Food Fishes,* 1921, p. 31.

45. Glen C. Leach, *Propagation and Distribution of Food Fishes, Fiscal Year 1923.* Bureau of Fisheries Document No. 964 (Washington: Government Printing Office, 1924).

46. Bower, p. 88.

47. H.B. Torrey and D.E. Lancefield, "Feeding of Raw vs Cooked Beef Liver," *Transactions of the American Fisheries Society,* Vol. XLIV, No. 2 (New York: American Fisheries Society, 1915), pp. 150-153.

48. Martin Norgore, "Salmon Eggs as Food for Salmon Fry," *Transactions of the American Fisheries Society,* Vol. LI, 1921-22 (New York: American Fisheries Society, 1922), pp. 82-85.

49. Bowen, 1970, p. 76.

50. Glen C. Leach, *Propagation and Distribution of Food Fishes, Fiscal Year 1924,* Bureau of Fisheries Document No. 978 (Washington: Government Printing Office, 1925).

51. E.M. Ball, "Inspection of the Karluk Hatchery," File "Correspondence concerning Collection of Salmon Scales, 1916-1918," Reports and Related Records, Division of Alaska Fisheries, RG 22, NA.

52. *Yes Bay Logbook, October 1, 1932-September 19, 1934* (Alaska Historical Library, Juneau, Alaska), April 22, 1933.

53. Ibid.

54. Kemmerich, July 30, 1979.

55. A.J. Spargue, *The Alaska Salmon and Their Practical Propagation* (Juneau: Alaska Daily Empire Print, 1921), p. 22.

56. U.S. Bureau of Fisheries, *Report of the United States Commissioner of Fisheries for the Fiscal Year 1922 with Appendixes,* Bureau of Fisheries Document No. 913 (Washington: Government Printing Office, 1923).

57. *Yes Bay Logbook, January 1920 to May 31, 1924* (Alaska Historical Library, Juneau, Alaska), May 17, 1920.

58. Leach, *Propagation and Distribution of Food Fishes, Fiscal Year 1923.*

59. Sprague, p. 21.

60. John Pease Babcock, "Some Experiments in the Burial of Salmon Eggs-Suggesting a New Method of Hatching Salmon and Trout," *Transactions of the American Fisheries Society, 1910* (Washington: American Fisheries Society, 1911), pp. 393-395.

61. F.M. Chamberlain, "Report of the Work in Bristol Bay and Elsewhere in Alaska," Reports and Related Records, Division of Alaska Fisheries, RG 22, NA.

62. A. Robertson, "Hatching Fry in Gravel," *Transaction of the American Fisheries Society,* 1919 (Washington: American Fisheries Society, 1920), pp. 145-156.

63. Leach, *Propagation and Distribution of Food Fishes, 1921.*

64. Kemmerich, July 30, 1979.

65. Letter from Harry D. Baer to Patricia Roppel, September 9, 1979.

66. Cobb, *Pacific Salmon Fisheries,* 1917, p. 222 states that 13,240,000 humpback salmon eggs were transferred from hatcheries on the Pacific Coast in 1913; 7,022,000 eggs in fall of 1914; and 7,000,000 eggs in the fall of 1915. These eggs were hatched and fry planted in various New England streams where conditions seemed favorable. Early in August, 1915, a female humpback salmon 22½ inches long and weighing 4 pounds, 3 ounces, was taken at the Bangor waterworks in the Penobscot River. Shortly after a male fish of about the same size was taken in this river at Orland Dam. A little later Bureau of Fisheries agents captured 20 alive near Bangor, and about 3,000 eggs were obtained and fertilized. In the Dennys River in Maine local fishermen caught a number of pink salmon in 1915.

67. "Materials Pertaining to Relations with Canada," RG 22, NA.

68. K.V. Aro, *Transfer of Eggs and Young of Pacific Salmon within British Columbia.* Fisheries and Marine Service Technical Report No. 861 (Nanaimo, British Columbia: Pacific Biological Station, 1979), p. 13-14.

69. Brice, p. 20.
O'Malley, *Artificial Propagation of the Salmon of the Pacific Coast,* pp. 23-28.

70. O'Malley, *Artificial Propagation of the Salmon for the Pacific Coast,* pp. 24-25.

71. Howard M. Kutchin, *Report on the Salmon Fisheries in Alaska — 1901,* (Washington: Government Printing Office, 1902), p. 68.

72. Ibid.

73. Howard M. Kutchin, *Report on the Salmon Fisheries in Alaska — 1905,* (Washington: Government Printing Office, 1906), p. 23.

74. John J. Coyle, "Report of John J. Coyle, 1902, Assistant Agent for Protection of Salmon Fisheries in Alaska," Reports and Related Records, Division of Alaska Fisheries, RG 22, NA.

75. Millard C. Marsh and John N. Cobb, *The Fisheries of Alaska in 1907,* Bureau of Fisheries Document No. 632 (Washington: Government Printing Office, 1909), p. 29.

76. Ibid.

77. A review on marking prior to 1904 can be found in "Early Salmon Marking Experiments of the Pacific Coast," *Pacific Fisherman,* Vol. II, No. 6, June 1904, p. 25.

78. Chamberlain, *Some Observations on Salmon and Trout in Alaska.*

79. Millard C. Marsh and John N. Cobb, *The Fisheries of Alaska in 1910,* Bureau of Fisheries Document No. 746 (Washington: Government Printing Office, 1911), p. 26.

80. Marsh and Cobb, *The Fisheries of Alaska in 1907,* p. 29.

81. Millard C. Marsh and John N. Cobb, *The Fisheries of Alaska in 1908,* Bureau of Fisheries Document No. 645 (Washington: Government Printing Office, 1910), pp. 57-58.

82. John P. Taylor, "Eighteen Months at the Karluk Hatchery," *Alaska Magazine,* January 1964, p. 36.

83. Charles H. Gilbert, "Age at Maturity of the Pacific Coast Salmon of the Genus Oncorhynchus," *Bulletin of the Bureau of Fisheries, Vol XXXII for 1912* (Washington: Government Printing Office, 1914) pp. 1-22.

84. John N. Cobb, "Report of Inspection of Alaska Fish Cultural Stations, Season of 1910," File "Private Salmon Hatcheries, 1905-1916," Reports and Related Records, Division of Alaska Fisheries, RG 22, NA.

85. Barton Warren Evermann, *Alaska Fisheries and Fur Industries in 1911,* Bureau of Fisheries Document No. 766 (Washington: Government Printing Office, 1912), p. 15.

86. Chamberlain, "Report of the Work in Bristol Bay and Elsewhere in Alaska."

87. Ibid.

88. Ibid.

89. Fred M. Chamberlain and Ward T. Bower, "Fishery Industries," *Fishery and Fur Industries in Alaska in 1912,* Bureau of Fisheries Document No. 780 (Washington: Government Printing Office, 1913), pp. 29-31.

90. Chamberlain, "Report of the Work in Bristol Bay and Elsewhere in Alaska."

91. Ibid.
Chamberlain, *Some Observations on Salmon and Trout in Alaska* states that "both dorsal fins" were removed which apparently meant that both fins on the dorsal of the fish were removed, i.e. the adipose and what today is known as the dorsal fin.

92. Chamberlain, "Report of the Work in Bristol Bay and Elsewhere in Alaska."

93. Ibid.

94. Ibid.

95. Ibid.

96. A letter from J.C. Barber of the Commercial Club of Ketchikan to Governor George A. Parks dated April 30, 1927 (Alaska State Archives, Juneau, Territorial Fish Commission File 12-10-1927) gives a description of the number of fry marked per day at the Ketchikan Territorial hatchery: "He [Wentworth] has succeeded in marking 3,000 fish in a single day but states that on account of the nervous strain of holding them and handling them without injury, this is much higher than the average work; 200 to 250 being a good average hour's work for a man."

III. The Voluntary Hatcheries

KARLUK RIVER FISHERIES — THE BEGINNINGS

The bountiful Karluk River was the logical place for the first Alaskan hatchery. There seemed to be an endless number of sockeye salmon swimming up river each year, and commercial fishing of this run began soon after the purchase of the territory from Russia. It was reported by 1892 that half the Alaskan production of canned salmon came from the Karluk estuary.[1] The geographic features of the Karluk River were conducive to successful fishing. The river which drains Karluk Lake, about sixteen and a half miles distant, flowed into a lagoon which was slightly affected by the tide. The lagoon was about 300 yards wide at its head and gradually widened until it was nearly a half mile across as it approached a narrow spit. Here the lagoon contracted and was at its outlet a minimum width of 90 feet (Map 4).

It was here on the spit, near the mouth of the lagoon, that cannery operations began. At least eight different canneries fished the Karluk River stocks during the 1890s. It was only natural that annual disputes arose over rights to this fishery resource. It was under these conditions that the first hatchery at Karluk was conceived. An apportionment of the fish was made for the year 1891 by mutual agreement of several canneries on the spit. This combine, Karluk River Fisheries, agreed that a hatchery might alleviate fear of a decrease in the number of sockeye salmon. The Karluk River Fisheries established an experimental hatchery on the lagoon about a half mile below the site of the future Alaska Packers Association (APA) hatchery and operated it one year.

Only about 500,000 of the 2,500,000 sockeye eggs incubated in the hatchery survived. This lack of success was said to be due to the water, obtained from a ravine, which became surcharged with impurities and covered the young fish with a scum or species of parasite.[2] Later analysis suggested that the causes were dirty water, unclean troughs and baskets, careless and insufficient picking of dead eggs, too sluggish current in the troughs, too small a supply of feed water, lack of aeration, and excessive water reuse.[3]

But fish cultural problems were not responsible for the closure of this first hatchery. It was the lapsing of the Karluk River Fisheries agreement after 1891 and renewed disputed between canning companies. Once closed, the hatchery building was offered to the U.S. Government. The U.S. Fish Commission, still in its infancy, did not elect to enter the Alaskan field at that time despite encouragement by the Commission's Alaska fisheries agents. One agent expressed the opinion that ". . . unless the Karluk hatchery is established, in addition to protective means, red salmon will be exterminated."[4] This argument was to no avail and the hatchery was abandoned.

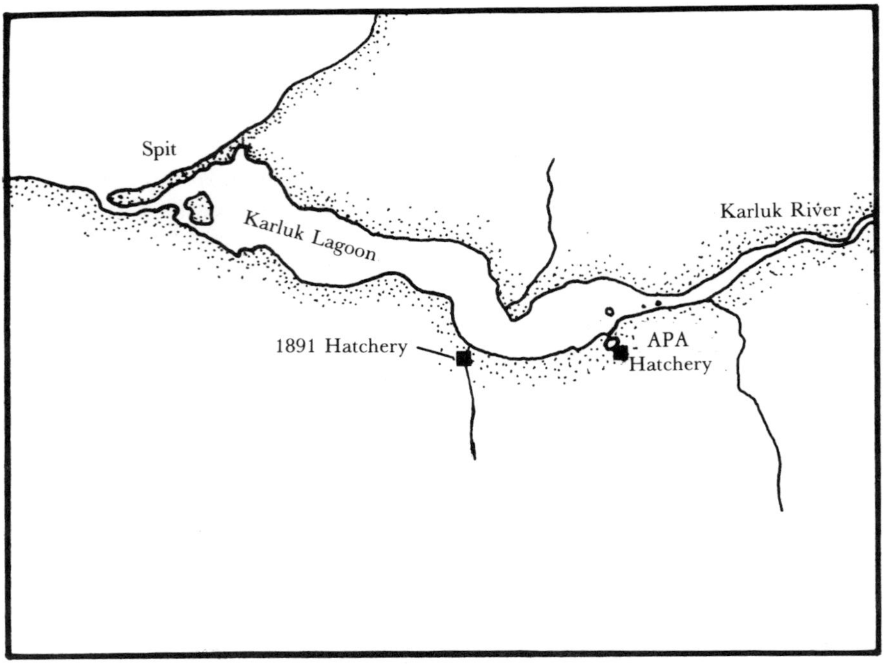

Map 4. Karluk Lagoon, Kodiak Island

JOHN C. CALLBREATH — PIONEER

Fish propagation began in Southeast Alaska a year after the closure of the Karluk hatchery. John C. Callbreath was one of the first to take positive action after he became convinced that the apparently inexhaustible salmon runs in Alaska could not last forever. His home base was Wrangell but his business, the Diamond C Trading Company, was located at Telegraph Creek, British Columbia. This Company constituted his main livelihood, except for a one-time participation in commercial fisheries. An enterprising individual, Callbreath encouraged Sanborn and Ellmore Company, an Oregon company, to establish a cannery in what is today the Bay of Pillars. Callbreath, as cannery manager, started the first Southeast Alaska hatchery in the spring of 1892.

The hatchery, generally referred to as the Point Ellis hatchery, was a primitive affair built on Katlaku (previously spelled Kutlakoo) Creek which empties into the Bay of Pillars. A dam was constructed across the stream, just above high tide and a person stood on the dam to dip the fish over the dam. The fish could spawn naturally, but the dam, Callbreath believed, kept out predators such as seagoing trout. A number of spawners were kept for the hatchery operations, and about a million sockeye eggs were incubated in outdoor troughs and baskets. An exceptionally high September tide destroyed the project.

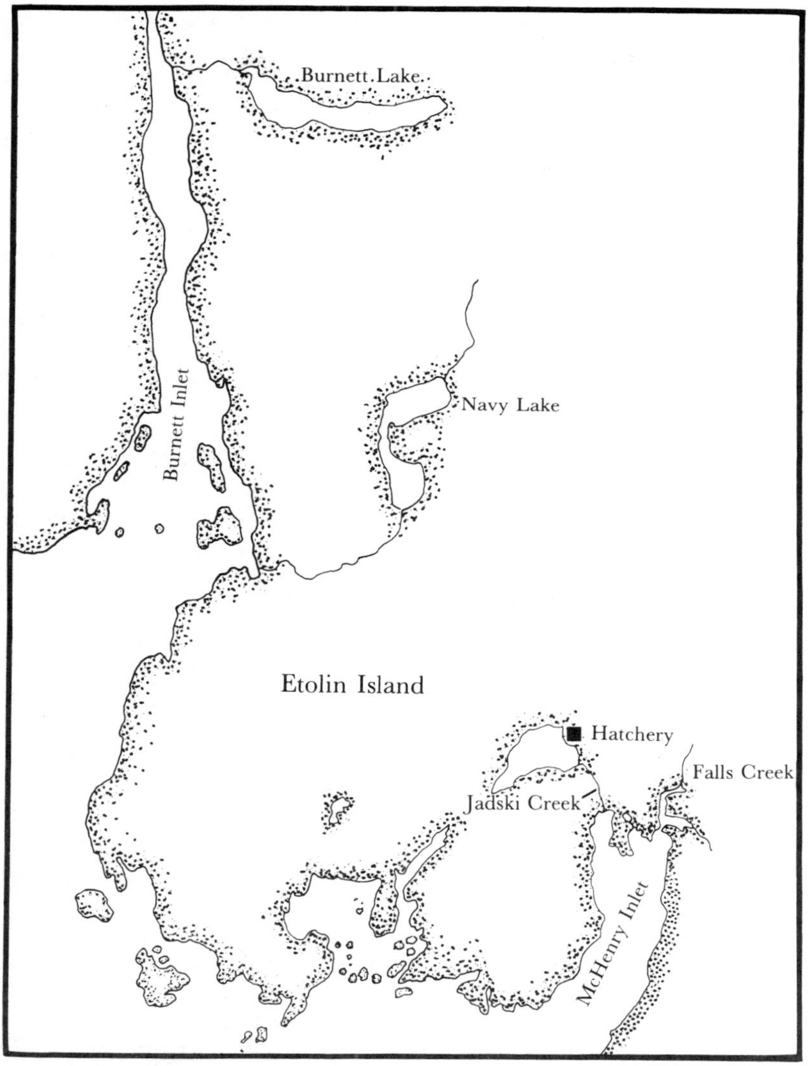

Map 5. McHenry Inlet, Burnett Inlet, and Callbreath's Hatchery Location

With the loss of the Point Ellis cannery by fire that same season, Callbreath ceased his commercial fisheries enterprise and turned his energies to another hatchery he had started. He had, in the spring of 1892, selected a site for a hatchery closer to Wrangell, and at the time of the Katlaku Creek flood, Callbreath was ready to take sockeye eggs at his new location.

This site was in a small bay in McHenry Inlet, Etolin Island (Map 5). The creek, a half mile in length, is known by the Indian name of Jadski Creek but called Jadjeska Creek by Callbreath. At the head of the creek is a small L-shaped lake, now known as Hatchery Lake, about three-quarters of a mile long with a surface area of approximately 500 acres. The system was never known to produce more than 3,000 to 5,000 sockeye and 1,500 coho.[5] This was a number too small in those days to attract attention of canning companies and the commercial fishermen.

After making arrangements with the Native who claimed the stream, Callbreath commenced work on a project to which he devoted many years, considerable effort, and large sums of money. Callbreath's plan was to select a stream which under normal conditions carried a few thousand sockeye salmon. By artificial propagation he hoped to increase the production to hundreds of thousands with the ultimate goal that, if he was successful, the government would enact a law making all the increased numbers of fish his property.

With no other hatchery in Alaska, Callbreath had no example to follow. The first fish culture station for Pacific salmon had started less than 20 years previously, and when Callbreath started his hatchery in Alaska, only two operated in California, three in Oregon and none in Washington. None of these propagated sockeye salmon.

Callbreath's knowledge of fish culture came from his conversations, visitations, and correspondence with fish culturists at many of these West Coast hatcheries. He patterned his operations after those of his friend Robert D. Hume and those of the Clackamas, Oregon hatchery. On his yearly trip to San Francisco, he generally visited Hume, who established a hatchery on the Rogue River, Oregon in 1877 and experimented with Pacific salmon culture. The Clackamas hatchery was built in 1876 by private enterprise and was later taken over by the U.S. Commission of Fish and Fisheries. Waldo Hubbard, superintendent, spent a few days in several different years with Callbreath and the two corresponded.

Many of his other correspondents kept him informed on fish culture activities not only on the West Coast but in Canada where a hatchery had been established in 1884 on the Fraser River. One correspondent kept him abreast of developments in France. He corresponded with David Starr Jordan, a leading ichthyologist, lunched with Jefferson Moser of the U.S. Commission of Fish and entertained most of the Alaska fisheries agents in his home in Wrangell.

Armed with knowledge from so many sources, his own resourcefulness, and trial and error experience, Callbreath began construction in 1892. He and a crew of two in 11 days built an eight foot dam across the creek and constructed a hand-sawn timber shack on the bank. Beside it they built six 16 foot and four 14 foot troughs and baskets (Fig. 12). The troughs had a capacity of 1,120,000 sockeye eggs and were covered but not enclosed. Callbreath took the first sockeye spawn in September and coho spawn in early October, 1892. Water temperature was 50 °F when the first eggs were set.

Figure 12. Troughs and baskets used by Callbreath in 1892, Jadski Creek, McHenry Inlet. *National Archives.*

Problems arose as would be expected in this pioneer stage of fish culture. Many of the adult salmon he had impounded in the stream to ripen died. The sockeye which were held in front of the dam ripened in salt water and Callbreath found the spawn worthless. The stream water, once incubation began, carried so much organic matter that slime accumulated, fungal growth began, and the eggs had to be picked daily. Eyed eggs developed in 32 days. "A great relief," Callbreath wrote in his diary. "Didn't see how I could possibly miss, but it worried me."[6] Hatching occurred between 90 and 100 days.

After this first year's experience with capturing spawners Callbreath was convinced that fish detained so close to salt water or short of the lake would not give perfect results. He searched for a new site along the lake shores, and located a warm, spring-fed stream. Air temperatures by mid-December had reached 20 °F so Callbreath immediately moved all baskets of eyed eggs and sac-fry to what he called Warm Creek. During the first winter the air temperatures reached 6 °F below zero, but the water in Warm Creek held at 39 °F. In subsequent years the lowest recorded temperature was 38 °F and the highest was 46 °F.

Callbreath moved his operation across the lake in 1893 where he could utilize the warm spring water. Lumber was shipped from Wrangell and packed to the site on the northern shore midway between a series of 11 springs and feeders. The hatchery, erected on the border of the lake, partly over Warm Creek, was 75 feet long by 11 feet wide with a shed roof and an open shed at one end (Fig. 13).

Troughs in the building were 16 feet by 13.5 inches by 5.5 inches (inside measurement) made of 1.5 inch planed lumber covered with asphaltum varnish. Two lines of 16 troughs, arranged in pairs with 8 troughs in a line, were in the main building. Each pair of troughs had a drop of one inch in its length, with a fall of four inches to the next. The first compartment in each trough was eight inches long and received aerated water. Next were seven basket divisions, each 24 inches long, separated by the Williamson system of division plates, two inches apart with the last space four inches. The shed would accommodate four additional troughs but seldom was used as winters were too cold.

Baskets, built by Callbreath, were of the usual wire webbing and had no wooden rims. They were supported an inch from the bottom of the troughs by broad-headed nails and tin clips. All were well lacquered.

The hatchery water came from a pool which was supplied by three springs in the immediate vicinity, augmented by an additional spring which was connected to the pool by a ditch. The water was conveyed by a covered flume 150 yards from the pool to the west end of the building. The flow was regulated at the closed end of the flume in the pool by means of holes in the bottom plank in which plugs could be inserted. The main flume was joined midway by another flume running from a reserve pool east of the hatchery. The water was clear and evidently pure as no trouble was experienced from fungal growth.[7] Filters were not used, but there were screens in the upper end of the flume to strain out particles which fell into the pool. The same water was used through four troughs and if the annex was used, through five. It then passed by a sluice to the small creek under the hatchery.

The method of handling spawning salmon was unlike that used in other Alaskan hatcheries. Callbreath believed that in order to obtain the best results only the most desirable species should be admitted to Hatchery Lake and that all predators should be removed and excluded. In order to accomplish this dams were built across the outlet stream with a rack below them at a point about 100 yards from salt water where an islet divided the stream. The dam and fence on the western side of the islet let nothing pass. The fence on the eastern side had a trap opening, admitting fish to the foot of the dam. Here the sockeye and coho were lifted by dip nets to the pool above, from which point they ascended quickly into the lake. No fish could enter the lake unless passed over the dam by hand and all undesirable species, the pink and chum salmon and Dolly Varden in particular, were carefully excluded.

Figure 13. Callbreath's hatchery, Hatchery Lake, McHenry Inlet, 1901. *National Archives.*

It was said that all fish such as cutthroat trout, sticklebacks, bullheads, and sculpins had been removed from the lake by fishing, but Callbreath believed these fish migrated and felt that they would be excluded when they attempted to return to the lake.[8] It is not certain he actually exterminated all fish in the lake. Pink salmon, which were excluded, were dipped from the trap in the dam, thrown into a box on shore, and left to die. Often the dead fish were returned to the stream. No pink salmon were allowed further than the barricade from 1892 until it was removed twenty years later. It was always a mystery to Callbreath how the returns of pink salmon continued despite his attempts at destruction of the species. It was not certain at that time whether pink salmon would spawn in the creek below the dam or perhaps in the intertidal zone.

The number of sockeye and coho salmon dipped over the dam was carefully noted in U.S. Fish Commission publications in the first years.[9] In later years only the number of sockeye was recorded (Table 5). Callbreath reported no change in the normal condition of his stream between 1892 and 1903 except that one-third grown sockeye, as he called them, showed up in later years in considerable numbers. Sixty-five ''jacks'' were taken in the stream in 1903, all sexually mature males.[10]

Table 5. Weir Counts, 1892-1912: Jadski Creek, McHenry Inlet

	Number of sockeye adults			Number of coho adults		
YEAR	Male	Female	Total	Male	Female	Total
1892	...	...	3,000	...	...	...
1893	...	...	3,010[a]	...	...	1,151
1894	2,438	2,016	4,454	256	230	486
1895	2,799	2,497	5,296	134	204	338
1896	1,617	2,008	3,625	374	338	712
1897	1,817	1,572	3,383	590	500	1,090
1898	1,189	821	2,010	158	142	300
1899	1,058	1,175	2,233	991	963	1,954
1900	1,991	1,863	3,854	526	482	1,008
1901	...	...	2,924	...	...	...
1902	...	...	5,077	...	...	...
1903	...	...	5,648	...	...	...
1904	...	...	7,432	...	...	...
1905	...	...	7,623	...	...	...
1906	...	...	7,320	...	...	...
1907	...	...	7,542	...	...	...
1908	1,022	876	1,898	...	...	...
1909	516	434	950	...	...	...
1910	...	...	...	...	...	...
1911	1,487	1,149	2,626	...	...	...
1912	1,282	1,153	2,435	...	...	...

[a] Weir not fish-tight for early part ot season.

Callbreath soon learned that once into the lake, the sockeye salmon were difficult to capture, and approximately one-third of the fish lifted over the dam were never recovered for artificial spawning. If the fish went into the small feeder streams to spawn, it was nearly impossible to secure them. As a result, Callbreath devised dams which he placed on six feeders and springs which formed the natural spawning beds. (Fig. 14). These dams had trap openings which admitted the fish, but did not permit them to leave.

The fish were dipped out of the ponds at spawning time. The wet method of spawning was used, and the eggs and milt allowed to set for one hour before washing. Callbreath stated he found that impregnation took place up to three minutes after the ova had been ejected and that best results were obtained by adding milt between one-half and one and one-half minutes after ejection.[11]

The carcasses were not always thrown in the lake. One of Callbreath's correspondents advised him not to throw the dead salmon in the lake because their decay destroyed the minute animalisms on which the young fish survived.[12] The carcasses were left to rot on the lake banks during the season of 1894. It is not known if this practice continued.

Incubation data are scanty. Eye spots generally appeared from 30 to 38 days. A few were earlier and a few 45 days before well eyed. The picking of dead eggs was done with tin forceps, the accepted tool of the time, and was started six to eight weeks after the eggs were placed in the baskets. In 90 days the eggs were hatching rapidly; in 100 days 67 percent were hatched; 110 days 80 percent were out; and the remainder straggled along for several months. The coho hatched about 10 days earlier. Callbreath closed the hatchery on March 1, and unhatched eggs were buried in gravel, simulating natural conditions. Combined coho and sockeye releases are shown in Table 6.

As the sac-fry collected in the lower end of the troughs, they were removed within a day or two of hatching to prevent loss by smothering. The sac-fry were placed in numerous nursery ponds and upper ponds of the feeders. All were natural springs which had been enlarged by dams, and some were the same ponds used to capture spawners (Fig. 14). Not all of the sac-fry were released directly into the ponds. In one nursery pond, sac-fry were placed under inverted troughs from the original hatchery on the creek. A wall of rock was piled around and over the troughs. It is not clear if this method was used each year.

During one of the first years of his operations, Callbreath dug up a portion of one of the natural spawning beds to see what the eggs were doing naturally. He found that the sac-fry took from three to four months to absorb the sac. By comparison in the hatchery situation the yolk-sac was absorbed in 45 to 50 days, but plainly showed at 60 days, though skinned over and in the belly. After that had taken place, most fry were taken from the ponds and placed in other feeders or in the lake. Fry were generally 1.5 inches long after four months. Losses, Callbreath told the U.S. Fish Commission, were 8 to 12 percent.[13] However he reported a loss of 15 percent in the nursery ponds alone in 1902.[14] This conflicts with statistics in Table 6.

In addition to normal fish culture activities, Callbreath experimented to determine the most successful methods of spawning salmon. He was working with salmon culture at a time when little was known about the limitations of artificial spawning. His curiosity was great. Because the sex ratio was often uneven, Callbreath experimented with hermetically sealed milt in 1893. He observed that the eggs fertilized with such milt, set September 20, were mostly eyed out but not hatched by January 8. "This is an item for the experts to solve why the bottled milt fertilized eggs are much longer coming out."[15]

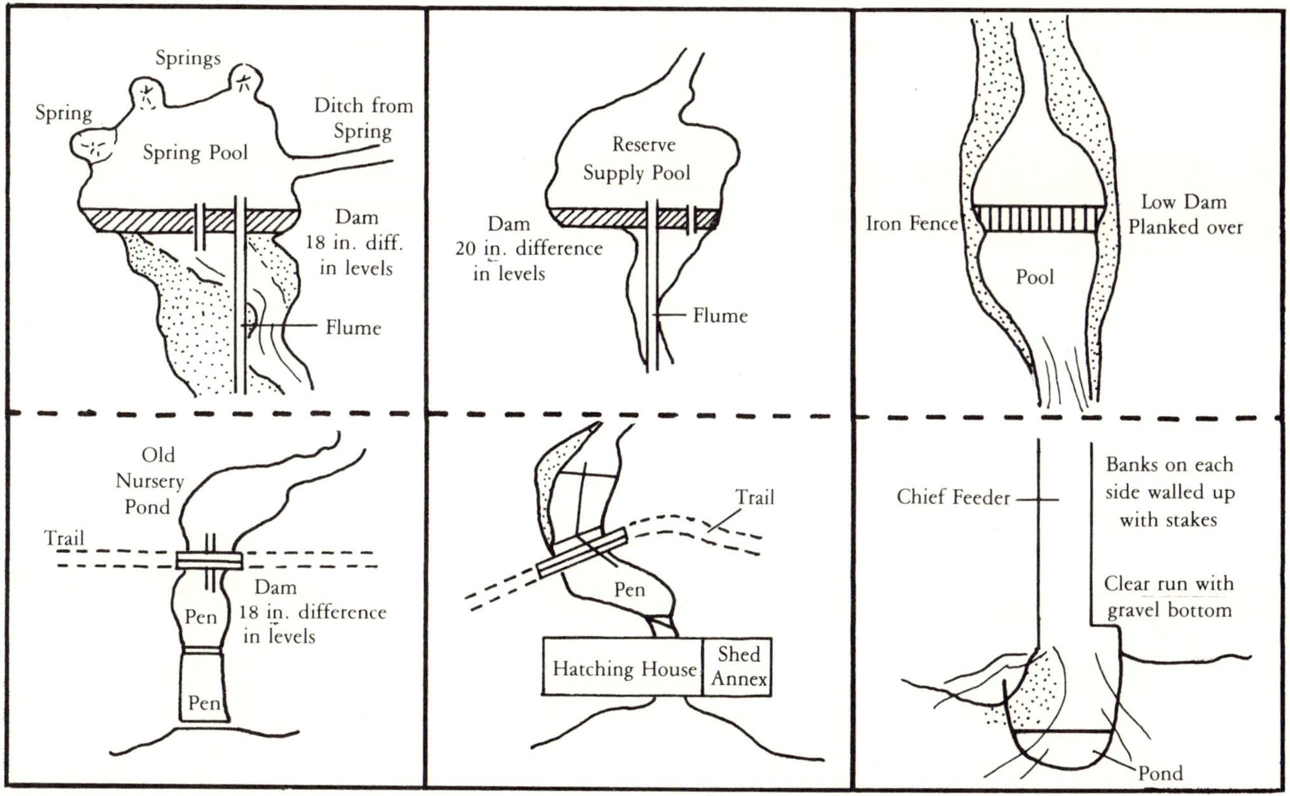

Figure 14. Diagrams of feeder streams and springs with traps for adult capture and for fry rearing, Callbreath's hatchery, 1901. *U.S. Fish Commission.*

Table 6. Combined Sockeye/Coho Production, 1892-1905: Callbreath's Hatchery

Brood Year	Number of Eggs Taken	Number of Sac-fry Released	Comments
1892	900,000	600,000	Sockeye eggs taken September 26-October 6.
1893	3,231,000	2,204,000	885,500 coho eggs taken — September 19-November 11. Coho sac-fry transferred.
1894	6,300,000	5,291,000	Sockeye and coho sac-fry transferred.
1895	6,200,000	5,475,000	Sockeye sac-fry transferred. 1895 diary gives 5,380,000 sac-fry released.
1896	5,400,000	4,390,000	Sockeye and coho sac-fry transferred.
1897	3,400,000	2,526,000	...
1898	3,000,000	2,050,000	...
1899	3,400,000	2,335,000	Coho sac-fry transferred.
1900	HATCHERY NOT OPERATED		
1901	6,000,000	5,500,000	Includes 4,420,000 Salmon Bay sockeye eggs. Sockeye sac-fry transferred.
1902	6,000,000	5,000,000	...
1903	6,000,000	5,000,000	...
1904	6,050,000	5,000,000	Includes 750,000 coho sac-fry.
1905	7,700,000	6,500,000	...

Callbreath tried other experiments with bottled milt, particularly during the year 1895. Sixteen hours after the parent fish was decapitated, 60,000 eggs were taken and fertilized with milt, 16 hours old, which had been kept in a dark place at an even temperature of 48°F in a tightly corked bottle. Fifty percent hatched as healthy sac-fry. He repeated the experiment, using 3,000 eggs and six hour old milt and still found a 50 percent survival. When he took 60,000 eggs from females as above, and fertilized them with milt from live males, he obtained results as he would normally expect.[16]

Callbreath devised another series of experiments to test more normal situations. Thirty-two thousand eggs were fertilized in a pan with five drops of milt and the spawn stirred well. He found that only 128 dead eggs resulted. About

3,000 eggs were washed before fertilization, and nearly all proved bad. Similarly 3,000 eggs were left standing one hour before fertilization and only one percent hatched. Another pan of 3,000 eggs was left eight days without change of water after fertilization at 33 °F and all were dead. When he repeated the experiment for five days, two-thirds were dead.[17]

Incubation procedures were only vaguely defined for sockeye salmon and Callbreath conducted experiments into this stage of the fish's development. A full basket of 60,000 eggs was not touched nor picked for 40 days. At that time the eggs were agitated by churning sufficiently to get at and break up the clusters. No dead eggs were removed. This process was repeated at intervals of 15 days and 85 percent healthy sac-fry were hatched.[18] Callbreath, however, did not adopt this method but continued to pick eggs daily.

Once the sac-fry were hatched, Callbreath looked for ways to relieve the overcrowding that soon developed in the troughs. He took 100,000 sac-fry in 1894 and put them in baskets through holes in the ice at the mouth of the lake. The fish were covered with bark and the hole covered with brush. "I do this," he wrote in his daily journal, "to see if they will survive in the cold lake water."[19]

Ten days later he uncovered a hole where 7,500 coho sac-fry had been placed and found ". . . the greater part of them dead, piled up in the basket, just as we had emptied them in. They had smothered in the basket."[20] He turned out the live ones. What became of the other baskets he did not record nor did he try this procedure again.

Another of Callbreath's experiments involved placement of sac-fry in barren lakes. He wrote in 1895 of this experiment:

> "I last year commenced an experimental process, which I feel encouraged to believe will prove successful; that is, by turning out a portion of my young fish in streams, on which there are lakes that fall into the sea by falls, over which no fish can pass. As a consequence, most of these streams and lakes are entirely barren of fish of any kind. There are three streams and lakes of this description contiguous to my hatchery. In the winter of 1894-95 I placed 1,000,000 young fry into one of these lakes, and the present season of 1895-96 will place 2,000,000 in another lake, and so keep on alternating until I prove whether they will return to these streams. At the same time I will keep on stocking my own hatchery lake with as many as I think it will sustain. If my experiment of stocking these heretofore barren lakes and streams proves successful, and I can see no reason why it should not, it will prove of great value to the salmon fisheries of this coast, as these lakes abound all along the Alaskan coast."[21]

Table 7. Egg and Sac-fry Transplants, 1893-1904: Callbreath's Hatchery

Brood Year	Species	Number of Sac-fry	Site Planted	Comments
1893	coho	66,000	Falls Creek	...
1894	coho	50,000	Falls Creek	...
1894	sockeye	1,000,000	Burnett Lake	...
1895	sockeye	2,000,000	Navy Lake	1,700,000 on Jan. 17, (1896); 150,000 on Feb. 7; 150,000 on Feb. 10.
1896	sockeye	1,000,000	Burnett Lake	...
1896	coho	135,000	Falls Creek	...
1899	coho	60,000	Falls Creek	...
1901	sockeye[a]			
		4,420,000	Cooney Creek Mosman Creek Hatchery Lake	2,500,000 sac-fry 500,000 sac-fry Reminder of sac-fry
1902	sockeye	6,000	Wrangell Narrows	Purchased by M. Kildall
1904	unknown	1,000,000	Falls Creek	Transplanted before Jan. 11, 1905 and on Jan. 13.

[a] spawn taken at Salmon Bay.

The lakes in nearby Burnett Inlet which Callbreath chose were Burnett Lake and what he called Francis Lake which is today's Navy Lake. In addition to these transplants, he also put coho sac-fry in Falls Creek, maybe as high as above the fourth fall (Map 5). A summary of these transplants is shown in Table 7.

Sac-fry were transported in coal oil cans as soon after hatching as weather permitted. A screw-top mouthpiece, 1.25 inches across, was soldered to the top of the can, and the fry poured in through a funnel. A five gallon can held 30,000 sockeye or 15,000 coho and two such cans, carried on the back with pack straps, made a load for one man. If the weather was cold, the cans were wrapped in blankets.

Callbreath released the majority of the fish in Hatchery Lake, except for his barren lake experiments. Economics and the mandatory hatchery act caused him to do otherwise in 1901. Both Wrangell area canneries, one at Point Gerard and the other at Point Highfield, contracted with Callbreath to produce the fry required by the new law. Callbreath stripped 1,600 sockeye females at Salmon Bay in the fall of 1901. These were taken by seining in the tidal basin as well as at a barricade. Approximately 4,500,000 eggs were taken.

This was Callbreath's first experience moving eggs from a remote site to his hatchery. Sometimes he had to wait two to three days for a steamer. Delay in moving eggs to the hatchery caused him many problems. Considerable losses occurred at first due to smothering in transport cans. Callbreath solved this by designing a small flume in the barricade. As egg buckets were filled, they were placed under the flume, and an auger hole bored in the bottom of the latter allowed a small flow of water to drop a short distance into the mouth of a funnel fixed to the inner side of the bucket. The drop ensured aeration. As the tube of the funnel passed to the bottom of the bucket, circulation was upward through the eggs.

The resultant sac-fry from the Salmon Bay spawn were planted in Cooney and Mosman creeks on Etolin Island. Callbreath was paid $2.50 per 1,000 fry by the packers, and he liberated 3,160,000 for which he, therefore, received $7,900. In 1902 he received an order from Pacific Coast and Norway Packing Company in Petersburg for 600,000 fry which he undoubtedly provided, but no details are available. That same year M. Kildall, a salter in Wrangell Narrows, reportedly received 6,000 sac-fry for his stream.[22] About that time many of the cannery operators began to ignore the mandatory hatchery requirement and Callbreath's services were no longer needed.

For all his interest in fish culture, Callbreath was most interested in the results of his labors. He was sure that his fish would return. He made this statement at a time when no one had demonstrated that salmon returned to the parent stream. Callbreath was taking a great chance in attempting to build sockeye runs to streams and lakes where few or none had previously been, so the U.S. Fish Commission watched this project with great interest.

Another unknown was how many of the fry would survive in the ocean to return as adults. Callbreath believed that a majority would return. Because of this belief, he advocated the granting of special privileges to those engaged in propagation of salmon. He requested the government guarantee continuous possession of the stream and adjacent water to a hatchery and a just and equitable share of the fish produced.

Callbreath felt his first returns would be in 1896 from 1893 releases. This was in keeping with what scientists generally accepted as the normal time for sockeye to return. In mid-July, Callbreath arrived at McHenry Inlet to find the bay full of fish, both at the mouth of Jadski Creek and Falls Creek. "Judge there must be 50,000 so that settles the hatchery question. It is a success beyond question."[23] The next day he cruised around the bay to see the extent of what he considered his return "and was much elated at the result, am sure there is from 50 to 100,000 in the bay now."[24]

The following day Callbreath returned to Wrangell to arrange for a boat to harvest the fish. When none was available, he traveled by steamer to Juneau to charter a boat. It was August 10 by the time he returned to McHenry Inlet. The bay was full of humpback salmon, but the sockeye were gone.

A haul was made, but it was so late in the evening that the men could not raise the net, so it was left until morning. Dogfish cut the net badly and destroyed many of the fish. Only 1,400 sockeye were taken. Added to the weir count, only 5,025 sockeye could be accounted for.

Over the succeeding years, Callbreath waited for the gigantic run of sockeye salmon to materialize. Because he believed so strongly that his fish would return, Callbreath began to expand the number of years it took for sockeye salmon to return to the home stream. He hypothesized 8 years, then 11 years, and finally 25 years. To those who are fully cognizant of the four and five year cycle of sockeye, his theories sound ridiculous. Yet, in the late 1890s and early 1900s no one could prove that Callbreath might not be right. Chamberlain's marking studies to prove the time it took sockeye salmon to reach maturity were still in progress at the time Callbreath closed operations.

Since so many people were watching Callbreath's operation, many theories were advanced to explain why the runs at McHenry Inlet did not increase dramatically. A fisheries agent wrote in 1900: "Many people believe that he [Callbreath] has not located on a good salmon stream and that his fish are going into other waters and that other men are reaping the reward of his labors."[25]

Moser reported in 1897 that there was a dramatic increase in the fish runs at Hetta, and the fishermen claimed that these streams were receiving adult fish produced at Callbreath's hatchery.[26] Callbreath answered: "It is not likely that my fish would make the circuit of Prince of Wales Island to select that stream, even if they were dissatisfied with their own stream, and pass by dozens of streams equally suited for their purpose."[27] He attributed the large catch to the fact that instead of one drag seine, ten or fifteen seines, some of them purse seines, were in operation at Hetta.

By 1901 Callbreath was aware of the fishing threat to his stream and the fish which returned to spawn. He began to complain to the U.S. Department of the Treasury, then in charge of Alaskan fisheries. "The canneries are all around me, and I have already had notice that my stream will be fished for red fish. One day's fishing will clean me out."[28] Four new canneries were being built within 75 miles of Wrangell and his hatchery. Callbreath asked that his stream be protected. After public notice and a hearing in 1901, Jadski Creek was declared a temporary salmon reservation. A second hearing was held in 1902 to continue this protection.

The largest return to Callbreath's stream appears to have been in 1905. That year sockeye salmon returned earlier than usual, and by August it was reported McHenry Inlet was full of salmon. A group of Wrangell men investigated and told the *Alaska Sentinel*, Wrangell's newspaper, on August 31, 1905 that there was an increase in salmon in the bay, and they were satisfied these were fish from the hatchery. The run was not sufficient to warrant special efforts to harvest them. Over 7,600 sockeye salmon were put over the dam into a system which had previously been considered to produce only 3,000 to 5,000 sockeye (Table 5). Callbreath was unable to see these fish. Cataracts, though operated upon, caused him to become blind that year.

Spawn was taken in 1905, but Callbreath was unable to maintain the establishment. Because he was not engaged in the canning or salting of salmon he was unable to secure rebates on the fish he released. The hatchery was operated for 14 years with the only remuneration for work or expenses occurring during the two years he sold fry to canneries. Callbreath continued to hire a man to put fish over the dam, but some were undoubtedly sold. The man when challenged, insisted he sold only pink salmon

Chamberlain, of the Bureau of Fisheries, discovered in 1912 that the dam was still in place and pink salmon were still being destroyed. "Just how this nefarious practice has evaded the law against wanton waste of food fish since the 1906 Act, passes understanding, nor is it clear under what authority this stream is barricaded since the hatchery was discontinued."[29]

The barricade was removed and with it ended one of the most unique artificial propagation attempts in Alaska. As a Bureau of Fisheries agent stated:

> "The efforts of Capt. Callbreath affords a most interesting sidelight in the history of Alaska salmon fisheries, and since their value from a scientific viewpoint is considerable, it is the more unfortunate that the heavy expenditure of time and money involved should have been proportionately so unprofitable in a practical way."[30]

CONTROVERSIAL KARLUK

The most well-known and extremely controversial of Alaska's salmon hatcheries was operated by Alaska Packers Association (APA) at the Karluk River on Kodiak Island. While the years of its operation paralleled the development of scientific knowledge of the sockeye salmon's life history, fish culture techniques at Karluk did not keep pace with these discoveries. Yet the politically powerful company was able to continue operations long after the Bureau of Fisheries knew the effectiveness of the Karluk hatchery was nil.

Its ineffectiveness was due to the selection of a totally inadequate site at the mouth of Karluk River. A precedent had been set in 1891 when the first hatchery had been built near the stream mouth. No criticism was forthcoming at either time by the Bureau of Fisheries because little specific fish culture knowledge was available in the 1890s and none was directly applicable to production of sockeye fry. The location of a salmon hatchery on a brackish lagoon did not appear to be impractical in 1896. The APA situated its hatchery on the southern bank of the Karluk River three miles up the river at the head of tidewater (Map 4). Ground was broken on May 28, 1896. The man in charge, who designed the layout, supervised construction, and then became superintendent, was James A. Richardson. He had formerly been with the U.S. Fish Commission at the Baird, California hatchery and had been trained by Livingston Stone, who was considered the father of salmon fish culture.[31]

Construction at the Karluk hatchery was so far advanced that August 29, 1896 Richardson and his crew began taking spawn which resulted in a season's take of 3,236,000 sockeye eggs. Thus, the hatchery started down its 20 year path of controversy, but during the first four or five years it was considered a model — well equipped and scientifically conducted.[32] Every effort was made to incorporate technology being developed on the West Coast at the turn of the century.

An efficient water supply system was designed to prevent the water problems which might have occurred at the earlier Karluk hatchery. Two water sources were developed (Fig. 15). One was Shasta Creek, a mile long stream entering the lagoon from the hills to the south. The other, called the ditch, came in from the eastward, carrying water from a small creek flowing down the side of Sugarloaf Peak. At the end of the ditch, its waters were received by a shed-covered tank known as the tank-house. Excess water from the tank-house found its way into Shasta Creek and a branch of the system ran a Pelton wheel to generate electricity. Part of the Shasta Creek water went into a settling pond which also received water from a small spring running year around. Excess water from Shasta Creek was used in a series of terraced spawner ripening ponds.

The supply of water to the hatchery itself was received from the ditch, via the tank-house, and from a flume from the settling pond. This settling pond was fairly deep and received its feed so gently it did not stir the sediment from the bottom or roil the water. The coarser debris such as straw, twigs and leaves carried from the creek, settled out before the water passed from the flume. Once into the upper level of the hatchery, the water passed through a filter system.

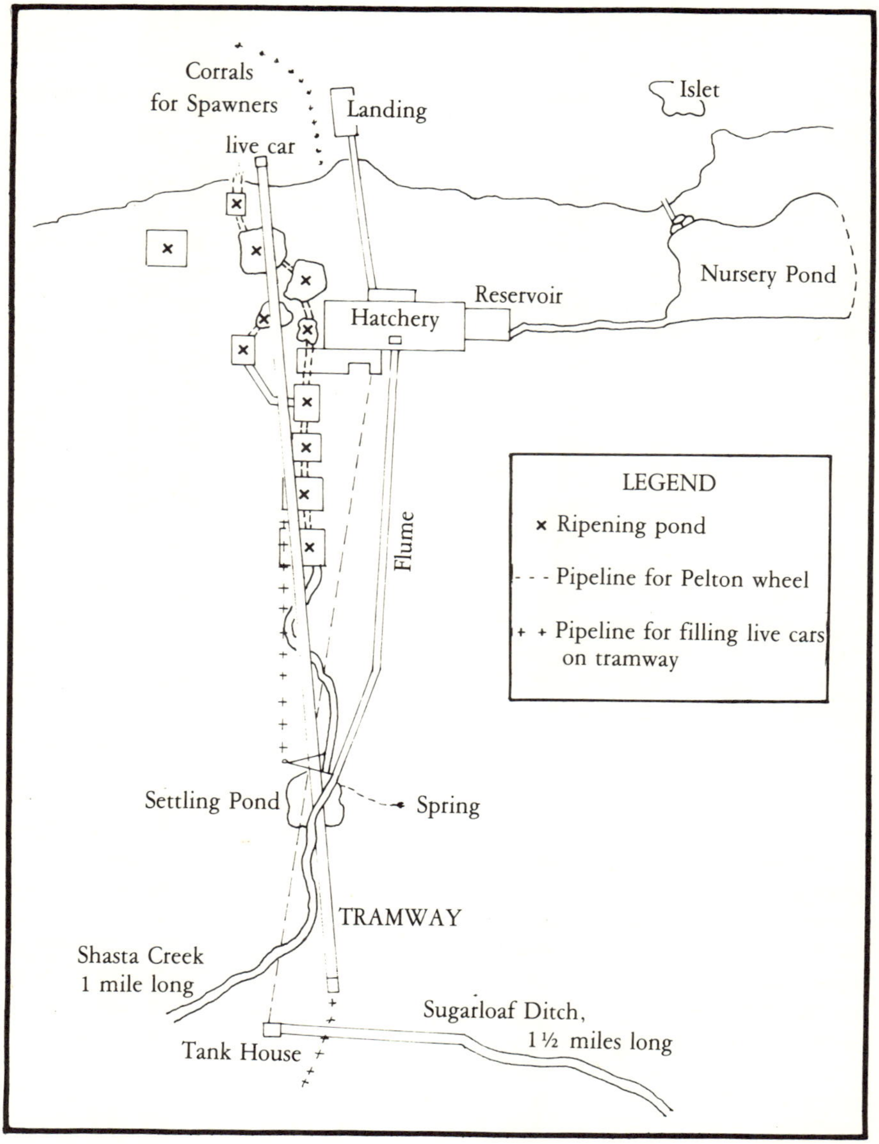

Figure 15. Sketch plan for the Karluk Hatchery water supply system showing sources of water and transport system for hauling live fish to the series of ripening ponds.

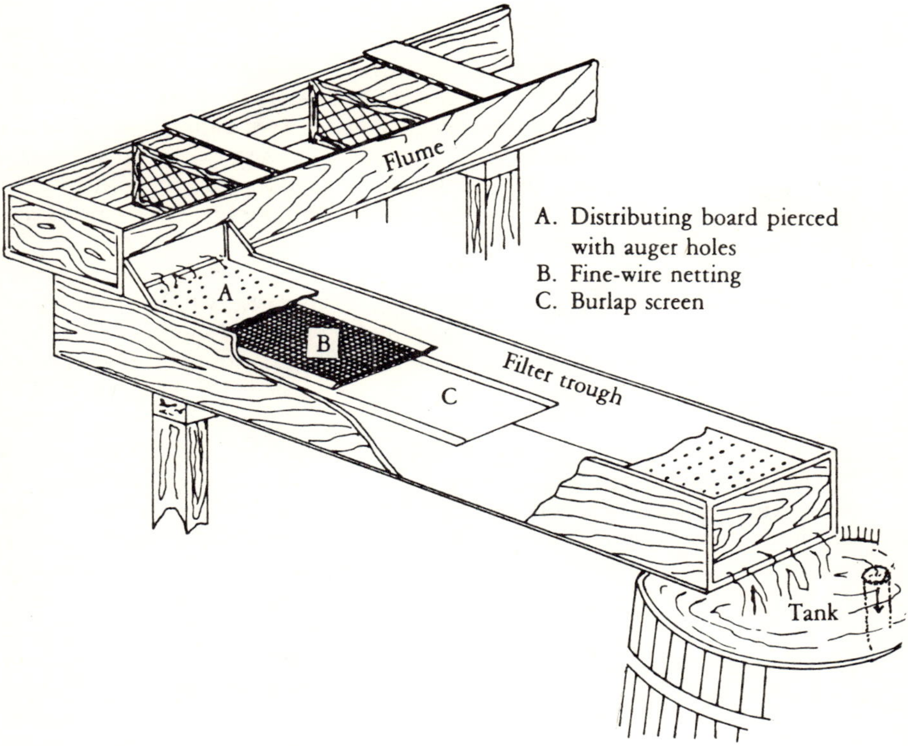

Figure 16. Richardson water filter developed at Karluk hatchery, circa 1900. *U.S. Fish Commission.*

The filter system was called the Richardson filter, apparently designed specifically for the Karluk hatchery by the superintendent (Fig. 16). The filter was a set of screens of varying degrees of fineness arranged in a long wooden trough. The water fell from the flume or pipe upon the upper end of the top screen which was a long board pierced with numerous auger holes. The holes increased in diameter as the farther end was approached, and the board had sufficient pitch to cause the water to run down and cover it. The function of the board was primarily distribution and incidently to aerate. Falling next through a fine mesh screen of wire netting, the water was received in spray-like condition by the last and lowest tray and screen. This was made of burlap, tightly stretched over a strong, wooden frame.

From the filter the water passed into a tank from which it was piped to the troughs. The waste water discharged into a reservoir and from it escaped through a narrow ditch into the nursery or rearing pond where the overflow passed through a short flume into the lagoon.

The incubation troughs were housed in the main hatchery building which was 32 feet by 100 feet. The interior floor arrangement in 1900 is shown in Figure 17. There were 12 sets of four Williamson troughs and one set of three giving a total of 51 troughs, all measuring 14 feet in length, 16 inches in width, and 7 inches in depth (inside measurements). They were built of two inch redwood. Each trough held five baskets, but only four were placed in a trough. The troughs were reportedly cleaned frequently and, thus having a fifth space unoccupied, the entire trough could be scrubbed without lifting a single basket of eggs.

A new section was added to the building during 1903 which contained an additional 52 troughs, bringing the capacity from about 23 million eggs to 48 million eggs (Fig. 18). The water system, except for new flumes, remained the same.

One of the major causes of concern about the Karluk hatchery operations was the method of capturing and holding adult spawners. Alaskan fisheries agent after agent described in horror the enormous mortality rate among these fish as they were being held to ripen. Because of the location of the hatchery at the head of tidewater, it was necessary to capture the spawners before they began to ascend the Karluk River to the lake and to hold them until ripe.

Methods of adult capture varied. The first year Superintendent Richardson put a net across a narrow part of the river and caught 2,000 fish. But before he could secure the fish he needed, the superintendent of a rival cannery demolished the net fence. The next year a new problem arose. It was found that the cannery fishermen were making frequent hauls within the lagoon, taking 8,000 to 10,000 fish daily. The fish escaping the seines of the spit were almost certain of capture inside the spit. Finally, when only an occasional fish reached the hatchery site, the hatchery crew joined the commercial fishermen on the outside of the spit.

After that year, 1897, at the recommendation of Jefferson Moser of the U.S. Fish Commission, the U.S. Treasury Department on May 7, 1898 prohibited fishing for salmon in the Karluk Lagoon except for the hatchery stock and for domestic use by Natives. The adult salmon, thereafter, were secured by seining crews under orders of the superintendent of the APA canneries at Karluk Spit. The principal seining grounds were the northern shores of the lagoon, opposite the hatchery at the mouth of the river.

The live fish captured in a seine haul were brailed into two live cars made of old dories with square ports cut between each frame over which galvanized wire netting was stretched. One fisheries agent felt part of the heavy loss was due to overcrowding in these dories.[33] In the early years the fish cars had no covers, and the fish thrashed about wildly. These live cars were towed to two corrals where the fish were tallied as they were dipped into the enclosure.

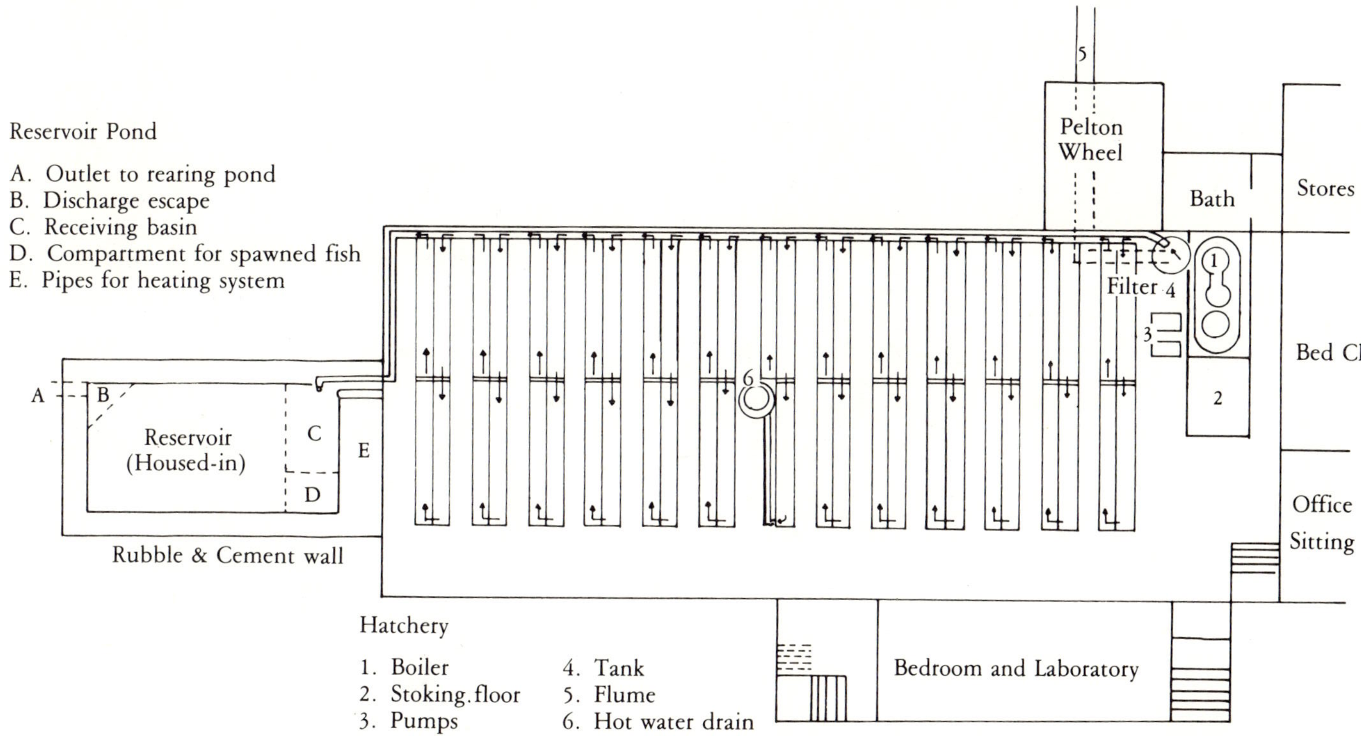

Figure 17. The interior floor arrangement, showing the layout of troughs and the water system at Karluk Hatchery in 1900.

Figure 18. Karluk hatchery complex, June, 1906. *Cobb collection, University of Washington.*

Figure 19. Corrals used to hold spawners in Karluk Lagoon, also showing eel grass across the lagoon where fry were often released, 1900. *U.S. Fish Commission.*

Figure 20. Freshwater ripening ponds, Karluk hatchery, 1900.
U.S. Fish Commission.

The two corrals covered an area of three acres and were normally in fresh water except at extreme high tide (Fig. 19). At high tide there was ten feet of water at the outer edge of the corrals and at the shore edge two to three feet. Rise and fall at extreme range was five feet. The corral fences were wire and cotton netting stretched between piles and stands of old iron pipe. It was reported that it was better, especially early in the season, to impound the brood stock in a large enclosure instead of holding them in small ripening ponds where, in a restless condition, the fish tended to become scarred and bruised causing fungoid growths.[34] Yet mortality of the spawners was great in the corrals. This was probably due to the rough treatment in the process of capture, one fisheries agent felt.[35] Another agent questioned whether the corrals were of sufficient size for the number of fish put in them.[36]

Fish did not complete their ripening in these corrals. Realizing the need for the sockeye salmon to complete its maturity in fresh water, Superintendent Richardson designed a series of holding ponds which terraced down the hillside. These were rectangular in shape and were walled with rubble and concrete so no mud would be stirred up by frequent violent movements of the fish (Fig. 20). The floors were of broken rock and gravel. The upper ponds were designed with sufficient fall between for aeration, but the three lower ponds lacked adequate aeration as early as 1900 and over the years nothing was done to correct the situation.

Again a seine was used to take the fish from the corrals to the ripening ponds. This 12 fathom seine was hauled into the shoal water off the lower end of a tramway. Four slatted floating cribs were secured to the corkline outside, two for males and two for females. The fish were picked from the net, sorted, and, if too green, returned to the corral. When the cribs were filled, they were hauled alongside the tramway. Fish were dropped into a car filled with fresh water and then transported to the ripening ponds.

This method for securing and handling adult fish was not effective. When operations were examined in 1903 by Wisner for the Alaska Salmon Commission, he was appalled at the stupendous losses which occurred among adults which were held. His report showed that between 1900 and 1902 inclusively 41 percent of the fish captured died before spawning; 60 percent of the females were lost and 40 percent of the males. He calculated that at the rate of 3,500 eggs per female, 333,375,000 salmon eggs were destroyed from 1896 to 1902, a period of seven years.[37] Records have survived for the last ten years of the Karluk operations and these show that this heavy loss of spawners continued (Table 8).[38]

Enough fish survived for spawn to be taken annually. When this process began, the water in the ripening pond was lowered to a depth of 21 inches or knee deep. The waste gates were arranged so as to let the water fall to that depth and no less so the fish would not smother. A panel of flat fencing was lowered into the water at one end and pushed slowly to the other side of the pond. The fish were dipped up, examined and if found ripe, placed in a floating car made of slats. If still green, fish were freed in the pond behind the fence. The last fish, which did not ripen as late as November or December, were often released back into the Karluk River (Table 8). Fish spawned both in the corrals and in the ponds and it was estimated that as high as 15 percent were thus lost.[39]

Spawning procedures were those in use at salmon stations of the U.S. Fish Commission, especially at Baird, California. The incision method was in use by 1919. The number of eggs per basket varied each season depending upon the number of eggs taken, but eggs from 25 females was generally considered a full basket. Fassett found in 1910 that 80,000 eggs were in some baskets and as many as 103,000 eggs in others.[40]

Incubation procedures appear to be similar to those used at hatcheries of the times. After baskets of eggs were placed in the troughs, the baskets remained covered from two to three days. Then each day, the eggs were floated once, very gently, and the dead eggs and empties on top were removed using tin forceps.

One unique feature in incubation at Karluk hatchery was an attempt to heat the water to hasten egg development. When the hatchery was constructed in 1896, a furnace and pump were installed in the basement to heat water for a

Table 8. Sockeye Salmon Used and Wasted, 1906-1915: Karluk Hatchery

Brood Year	Adults Taken	Females Spawned	Males Spawned	Returned to River[a]	Fatalities	Eggs Taken
1906	80,347	13,037	11,120	19,594	36,596	38,679,200
1907	95,734	15,507	14,720	14,258	51,249	47,808,200
1908	71,320	14,074	12,588	9,135	35,523	40,320,000
1909	95,804	15,144	14,075	6,628	59,957	45,228,000
1910	85,623	17,881	17,390	8,178	42,174	49,626,000
1911	79,699	14,516	14,770	4,747[b]	30,786	41,026,800
1912	69,053	14,219	14,929	8,794	31,111	45,500,000
1913	62,507	11,138	11,997	5,149	34,223	34,629,160
1914	59,684	11,900	11,624	1,073	35,087	30,240,000
1915	87,091	15,698	16,098	4,673	50,622	41,135,000

(a) Fish which had not ripened by November or December were released.

(b) 25,424 fish escaped from Corral Number 2. There is a discrepancy in the total number of fish in original records.

Source: "Karluk Hatchery, 1903-1954," Auke Bay Fisheries Laboratory Records, National Marine Fisheries Service, Federal Records Center, Seattle.

Table 9. Incubation Data, 1906-1915: Karluk Hatchery

Brood	Avg. °F. to Eye			Avg. Days to Eye			Avg. °F to Hatch			Avg. Days to Hatch			Sac Absorption	
Year	First	Middle	Last	First	Middle	Last	First	Middle	Last	First	Middle	Last	°F.	Days
1906	45.5	43.5	37.5	30	51	81	39.5	37.0	35.0	145	176	231	35	90
1907	46.0	39.0	37.0	40	55	95	41.0	36.5	34.0	145	195	225	36	70
1908	46.5	42.0	38.5	38	50	85	43.0	36.0	33.5	125	180	240	35	80
1909	49.0	40.0	34.5	32	50	120	42.0	35.0	33.0	125	200	240	35	90
1910	48.0	39.0	35.0	34	54	120	42.0	38.0	33.0	125	180	250	35	90
1911	48.0	46.0	41.0	38	60	111	48.0	42.0	36.0	107	180	223	39	95
1912	48.0	48.0	47.0	41	49	97	49.0	47.0	44.0	104	155	212	46[a]	97
1913	51.0	42.0	40.0	40	51	78	41.0	40.0	37.0	97	155	181	37	80
1914	50.0	48.0	44.0	35	40	60	48.0	45.0	41.0	97	100	155	39	80
1915	52.0	50.0	38.0	36	42	86	50.0	40.0	37.0	83	111	185	43	55

[a] Possibly an error, as number was questioned by JTB on original copy.

SOURCE: "Karluk Hatchery, 1903-1954," Auke Bay Fisheries Laboratory Records, National Marine Fisheries Service, Federal Records Center, Seattle.

system of piping used solely for artificial heat. A set of coils was arranged directly under each section of troughs and in the bottom of the reservoir pond. When the artificial heating system was in use, warm water from the reservoir pond was pumped into the filter system upstairs. After running through the troughs the waste water was discharged back into the reservoir pond. It was necessary before using this warm water again to filter and aerate it. Some new water was added to freshen the supply without materially reducing the temperature.

If the water had not been used over and over again in this manner, it would have been impossible to heat it sufficiently during the cold winter months. Fassett, on his inspection trip in 1902 felt the system of aerating and freshening was so well controlled that the feed water remained good after repeated use.[41]

Incomplete data are available for 1896 when the heating system was in use. Fassett lamented that no exact data were kept on thermal conditions during the forcing process, but he was told in 1900 that the temperature was slowly raised 10 °F to 12 °F above the prevailing temperatures and kept at this point, about 45 °F to 48 °F.[42] Eggs taken August 29 eyed in 60 days and hatched in 165. Eggs taken September 16 eyed in 100 days and hatched also in 165 days with artificial heat used the last 42 days. Eggs taken September 23 eyed in 111 days. It is difficult to use these data for comparison with later years as shown in Table 9 because of the lack of dates on the latter information.

Fassett was told that no artificial heat was used during 1897. Whether the heating system was used more than once is unrecorded. It is probable the experiment was not repeated because one fisheries agent reported that "the greatest discouragement was the unexpected large expenditure of fuel required to keep the proper temperatures."[43]

Although records were poorly kept, some handwritten incubation data were found and are shown in Table 9. Two different runs of sockeye salmon were utilized, an early, which were spawned in June and July, and a late run in September and October. Data were calculated for an intermediate group, but no explanation accompanied the manuscript. The reliability of this information is unknown.

In 1902 Fassett made some observations on incubation. He was told that the eggs from the early run eyed more quickly (35.4 days over a ten year average) than the late run (93.3 days over the same ten year average). Hatching varied from 115.3 days for the early run to 214.2 days for the late run. Fassett felt that this was not entirely due to water temperatures because the early run seemed to have more vigorous vitality and hatched more rapidly under similar thermal conditions. "But this is still a matter of opinion and must remain so until the collection of sufficient data from which to draw careful conclusions."[44]

After the sac-fry hatched, they moved to the bottom of the trough, there being one-half inch clear space on either side of the basket and about the same beneath. Despite the fact that sac-fry required more space as they matured, they were usually held in the parent trough until swim-up, a period of about 11 weeks. This procedure was in use until 1903 when the egg capacity was doubled but the ability to hold sac-fry was not. After that, only a small number of sac-fry were kept in the troughs until June when the troughs had to be prepared for the new spawn. As the fry became free feeders in the troughs, they were allowed to escape via the wastewater pipe into the reservoir pond, thence through the ditch into the rearing pond.

Sac-fry were transferred to the rearing pond which covered approximately three-quarters of an acre. When the pond froze, holes were cut through the ice. Fish were poured from a bucket into a different part of the pond each time. It was assumed the sac-fry would work into the gravel and silt where they remained, for the most part covered, until the yolk sac was absorbed. Superintendent Richardson believed that the strong, free swimming fry craved salt water, and he usually kept a barrel containing a small amount of salt submerged in the pond. Whether this practice continued after 1906 when Richardson left is not recorded.

The fry were fed, sometimes in the troughs and sometimes in the rearing pond. The food most often used was tinned salmon or ''do-overs'' furnished by the canneries at Karluk Spit. The food was removed from the can, thoroughly desiccated, and then ground up in a fine-cutting sausage mill. Fry were reported not to thrive on this diet, refusing it after 10 days to 2 weeks.[45] Fisheries Agent Cobb in 1906 stated that the fry were being fed trimmings from the can filling machines, which had been chopped in a food chopper. Fry were devouring it greedily at the time of his visit.[46] After the fry were three to four months old, it was observed that they seemed to deteriorate in the rearing pond, and consequently, they were rarely held beyond this time.

The rearing pond was drained and cleaned twice a year, once early in May and again in September or October. Fassett was told this was necessary as a slime formed on the bottom and would get into the gills and smother the fish.[47]

It was the method of planting the fry that created the most criticism of the Karluk hatchery. The procedure used was dictated by the location of the facility. When the Karluk hatchery was located in 1896 there was no proven reason not to release the fry into brackish waters. When ready to release fish, the hatchery personnel drove the fry out of the rearing pond through a short flume into a live car made of an old skiff (Fig. 21). When the live car was loaded with 200,000 fish, the wire-screen gates were let down across the inner end of the flume to prevent escape of the remaining fry. Fry releases were then made at high water along the shores of the lagoon between the mouth of the river and the mouth of the lagoon. The location selected depended upon existing conditions such as ice, wave action, and movement of predatory species. The rocky shores opposite the hatchery and those a short distance below were favorite

Figure 21. Live car, made from old skiff, used to release fry. Fry transported to live car in flume, 1900. *U.S. Fish Commission.*

places for releases. Another was a large patch of eel grass within the upper lagoon in front of the corrals. This can be seen in Figure 19. When the pond and the upper part of the lagoon were frozen over, which occurred from the latter part of November to the middle of May, the fry were placed in tanks, transported on sleds to open water in the lagoon, and planted.

Once released, movement of the fry was observed. Some left the lagoon, and fry were seen along the outside beaches in the vicinity of the shaded rocky points.[48] These fry returned to the lagoon at frequent intervals, working in and out with the tides for a long time or until late fall. Local residents told Fassett that when the ice on the lagoon broke in the spring, the lagoon was clear of trout and salmon of any age.[49]

An unidentified man stated that he had seen windrows of dead fish on the beach near the Karluk hatchery. The superintendent of the APA cannery for the Karluk region offered a standing reward for anyone who could produce sockeye fry found on the Karluk Spit or on the beach in that region. No one claimed the reward. Dr. Barton Evermann, chief of the Alaska Fisheries Service, testified in Senate hearings that the chances were six to one that the fish originally seen were sand lances.[50]

Although the hatchery was operated by Superintendent Richardson with what are now considered unsound biological practices, Richardson showed a great interest in fish culture and displayed more curiosity into the life cycle of the sockeye salmon than many people of his time. He told Fisheries Agent Fassett in 1900 that:

> "They [adult sockeye salmon] put in an appearance as grilse, invariably males, which are thoroughly mature and that these grisle are two years old, that is from birth. Part of the breeds return in three years, while others of slower development return in four or five years, and it is not beyond the bounds of reason to presume that some may be delayed for an even longer period before reaching full maturity."[51]

Richardson carried on several marking programs during his tenure to prove his theory. On the 1897 brood the adipose fin was cut from 513 fry, part of which were liberated in the Karluk River and part in the lagoon. These fry had been reared to between one and three-quarters inches and four and one-half inches in length and were liberated in the fall. He claimed in the fall of 1901 that some marked salmon had been caught in the commercial catches.[52]

Each year Richardson continued to mark a few fry or fingerlings either by removal of the adipose or dorsal fin or notching one of the other fins. Mature fish were taken at the Karluk fishery showing these marks, he claimed, but the information conveyed was meager. The record of marks was so obscured by repetition of the same method during different years that it was impossible, even at that time, to form any conclusions.[53] As late as 1909 many fry were marked by clipping one or more fins according to one of the employees.[54]

Most official reports made during Richardson's tenure were favorable on conditions, procedures and cleanliness. Then, in 1906, he retired because of ill health and his replacement took charge in May, 1906. Ingwald Loe had first worked at the hatchery in 1903 as a carpenter. Loe's appointment must have come at nearly the same time as the certification for the hatchery made by the Secretary of Commerce and Labor under new regulations. This certificate was made without further inspection by federal fisheries agents all of whom were familiar with Richardson's operations. The only adverse comment on the certification was made by John N. Cobb who stated "It is my belief that better results could be obtained if it [the hatchery] were moved up to the lake. . . ."[55]

The site was remote, the few fisheries agents were overworked, and all past reports were satisfactory except for Wisner's 1903 complaint on the mortality of spawners in the ripening ponds. Loe operated the hatchery for four years without an evaluation of the operations.

In 1910 H.C. Fassett, who had made an extensive study and written description of the hatchery in 1900, again visited the hatchery. His report began a series of waves which culminated six years later in closure of the hatchery. Fish culture methods were not up-to-date, Fassett complained in 1910. Neither egg picking trays or salt baths were used "the superintendent being unacquainted with these two very admirable adjuncts to the fish-culturists' art."[56]

Again the topic of the ripening ponds and the subsequent mortality was discussed. Fassett did not dwell on the number of fish lost each year. This had been done by Wisner in 1903 and would be done again by Ball in 1916. Rather, Fassett concerned himself with the conditions leading to the mortality rate. He wrote:

> "All the lower ripening ponds are badly off for feed-water and the fatalities amongst the stock fish held in them are heavy. When removed to the upper ponds, where the water is fresh, the fish are found to improve; and the losses of fish placed in the upper ponds at the same time that others, of the same lot, are put in the lower ones, are very much less. The same water is used over and over, from pond to pond, without being freshened, and although the water is well aerated in falling from pond to pond, well designed apparatus for this purpose being used, it is clearly not sufficient to remedy the evil. The ripening ponds are very heavily stocked, unnecessarily as it seems to the writer, and the small flow of water allotted to the pond system must become very foul and vitiated before reaching the lower levels.
>
> "There is a pump in the hatchery, and an idle donkey engine by which it could be run, but when asked why he had not put them in use, the superintendent seemed to think it a matter of no great importance as there were plenty of fish to be had. The subject did not interest him very much, and he seemed greatly surprised when told that any waste of stock fish which could be remedied was wanton, and an unlawful practice. A look over the surroundings showed that it would be possible at little expense, to run a pipeline or box flume, around the hillside, taking water from the river a short distance above the hatchery, and this scheme was suggested to the superintendent for his consideration. Something of this sort should be done if the hatchery is to be continued at this present location.
>
> "Fish are not well protected against their own struggles to ascend the hillside in which the ponds are terraced, and they appear to be handled with unnecessary, avoidable, and regrettable roughness. There was an air of general unkemptness and neglect about the whole pond system, like a garden gone to seed, and a general lack of care was observable throughout the entire plant, although no important feature seemed to have been neglected. Carelessness and a lack of appreciation of the importance of close attention to minor details perhaps might best describe the situation." [57]

On the subject of fry releases into the lagoon, Fassett put into the official record for the first time the absolute necessity of the hatchery liberating fry into a lake system. Others before him had suggested this, but no one had gone so far as to insist. In his report, he said:

> "There can be little doubt . . . that practically all the red salmon fry resulting from the hatching operations of the Alaska Packers Association at Karluk have been allowed to go to sea within a few weeks of the time of their becoming free feeders. That this is not a natural condition with the red salmon of the region is well known, as is evidenced by the annual migration from the lake of fingerlings of at least a year's growth. Whether the tiny fry liberated in the lagoon are able to cope with conditions at sea at so tender of an age and state, and whether they are able to find sufficient proper food at this time, are matters which are not at present known. It seems reasonable, however, to suppose that the instinct which keeps the red salmon hatched in Karluk Lake within its confines for at least a year has some sound physical baisis. If the red salmon is a selective feeder, as is believed, and if the pursuit of that class of food upon which it depends for sustenance in saltwater, demands a physical vigor such as is not obtained without at least a year of growth, it does not seem that the chances of reaching maturity are good in the case of the Karluk hatchery product. For this reason alone, it is urgently recommended that every effort be made to induce the Alaska Packers Association to either move their hatchery to a place on the lake or that proper arrangements for planting the fry in the lakes be made.
>
> "By extending their nursery ponds so as to accommodate fry to the full capacity of the hatchery, which is believed could be easily and cheaply done, and by the construction of a light tramway by means of which to transport the fry to the lake, it is thought that the problem could be solved."[58]

Fassett went one step further by asking the Commissioner of Fisheries for disapproval of the certification required by the 1906 fisheries act.

> "It is urgently recommended, however, that after the closure of the spawning season of 1911, approval of the Karluk hatchery of the Alaska Packers Association be withheld until 1) better arrangements are made for care of the stock-fish held in the ripening ponds, 2) arrangements are made for planting or liberating the young fry in water other than brackish or salt, as in a lake, and 3) a permanent record, daily, of essential data is kept at the station, available for future study and reference."[59]

The Commissioner of Fisheries sent Fassett's report to APA with a request that the company consider the feasibility of releasing fry in the lake. It was said this would remove the doubt attached to the efficiency of methods being used. Furthermore, the Commissioner went on to say:

> ". . . at the time Karluk was approved it was realized that its planting methods introduced an unknown or doubtful element into the question of its efficiency, but it was not felt that the Bureau had information definite enought to justify withholding approval. Since that time there has been some addition to our knowledge of the residence of red salmon fry in fresh water, tending against the efficiency of planting at or close to the sea and a new hatchery in such a location would not be approved until it could provide especially a lake or proper fresh water planting. However, the Bureau is not of the opinion that brackish water planting is so discredited that approval of the Karluk hatchery should be revoked forthwith. . . .
>
> "The Bureau does not wish to take action in the matter of revoking approval until it can point to more definite facts or investigations than now exist as a basis of action."[60]

The APA retorted by giving a history of the hatchery and reminding the Commissioner that in earlier years it had received high commendation from government officials. The hatchery had been built at great expense on the only site that appeared feasible at the time and received no government recognition until 1906. The APA officials assured the Commissioner: "In view of the fact that better conditions probably prevail near the natural spawning ponds, we have had a reconnaissance made of Karluk Lake, intending at some future time, when our business warrants it, to build an additional hatchery near the lake."[61]

Jefferson Moser, who by this time worked for APA, went on to add that the company felt that, although they wasted some fish by holding them, such waste was not as great as in nature. He told of finding fish, full of spawn, dead under natural conditions along the stream banks while he was working for the U.S. Fish Commission.[62] Moser was a difficult person to refute as he was still considered an authority on Alaskan sockeye salmon. The Bureau ordered F.M. Chamberlain, who was then at Nushagak to substantiate Moser's stand.

It was so late in the season by the time Chamberlain arrived at Karluk, few dead fish were available for inspection for retained spawn so he inspected the hatchery. Chamberlain's report on the hatchery was favorable.[63] He found the retaining ponds still stocked and a few fish in the corrals. A few dead females were taken from the pond, but Chamberlain thought the loss was about that experienced in California hatcheries. The water system seemed satisfactory and the eggs were in good condition. He thought the equipment in the hatchery surpassed that of the government's Afognak hatchery since the Karluk facility had a large rearing pond and all sac-fry were retained until swim-up.

The 1910 controversy died after that, although APA replaced Loe, the carpenter-superintendent, with L.W. Strass until 1913 and put Edward Sargent, a government hatchery employee from Afognak, in charge that year.

Part of the reason the Bureau did not pursue the issue of Karluk's technology was that in 1912 a bill, S. 5856, was introduced into Congress which provided, among other things, that rebates for the operation of private hatcheries should be discontinued after the succeeding year. As a result of hearings a modified draft of the bill provided that all private hatcheries should be taken over by the federal government. Over the years these concepts were consistently proposed in fishery regulations affecting Alaska but nothing ever came of this proposal or any other legislation.

As a consequence of constant anticipation of federal acquisition of private hatcheries, nothing changed at Karluk. The Bureau fielded any inquires or criticisms with the statement that the Karluk hatchery was established under the auspices of the U.S. Department of Treasury. It was operated in good faith for ten years before any benefits in the way of rebates were received and, furthermore, as the Secretary of Commerce and Labor approved the hatchery in 1906, the Bureau felt a serious injustice would be done if it should withhold its approval of the hatchery as operated. It was true, the Bureau stated, that disapproval would have been given as early as 1912 if the legislation had not been introduced.[64]

Everything came to a head in 1916. Instrumental in the ultimate closure of the Karluk hatchery was a study which tended to prove the inefficiency of the hatchery. Finally the Bureau had an argument that APA could not refute.

The study was made by Dr. Charles H. Gilbert, who examined salmon scales collected in 1915 by APA at the Karluk hatchery. His findings tended to show that comparatively few of the salmon produced by the hatchery ever reached maturity. In order to pursue the matter still further, he requested scales from at least 400 sockeye adults spawned at the hatchery in 1916. As it turned out, the hatchery was closed by that time but the Bureau's agent took scales from sockeye salmon in the Karluk area. A copy of Gilbert's report has not been located, but his conclusions undoubtedly were reached in the same manner as a study he performed in British Columbia, the results of which were published in 1914. This study, premised on Gilbert's ability to read the age of the adult salmon from its scales, concluded that

> ". . . examination of the scales from all the important blueback [sockeye] streams of the Province [British Columbia] has shown for each basin that adult fish are derived from yearling migrants, to the practical exclusion of those which migrate as fry . . . . It would seem, then that with few exceptions the fry of this species perish after entering the seas."[65]

The next step leading to the closure of the Karluk hatchery came when E.M. Ball, the assistant fisheries agent for the Bureau, made a routine inspection of the Karluk hatchery to report the number of fry and eggs on hand, the number of fry planted, and the general condition of the work. He examined the books and records and found that the ". . . hatchery was temporarily in charge of Victor Carlson who assumed responsibility upon the death of superintendent Edward Sargent, late in March. While Mr. Carlson did what he could do to facilitate the inspection, he was considerably at sea as to the record of operations."[66]

After Ball counted the eggs in five baskets to find the average number, he determined there was a total of 5,340,000 eggs on hand. Carlson's records showed 8,334,000, a discrepancy of 2,933,420 eggs. Ball felt his calculations to be more reasonable: 72,170 eggs per basket compared to 112,486. When the calculation of fry on hand was made, Ball's figures were 9,6000,000 fry. The books carried 16,998,550, for a discrepancy of 7,398,550 fry.

The records showed a small loss of eggs and fry for the brood year 1915, but Ball questioned this after what he saw in the rearing pond:

> "The hatchery workmen made a practice of dumping all dead fry removed from the incubating troughs into the waste water trough where they are carried into the pond. In this manner, dead fry had accumulated until they literally covered the bottom of the pond and formed a layer that appeared to be several inches in depth near the intake flume, gradually decreasing in depth as it extended toward the edges of the pool."[67]

Ball was convinced that the records, despite Fassett's recommendations for improvement in 1910, continued to be inadequate and could not be trusted. He said of these reports:

> "The record of losses of both eggs and fry is indeed meager. No permanent daily record of either is kept. Losses day by day are noted on blackboards in the hatching rooms, and at the end of the month are entered as monthly totals in the superintendent's records kept at the hatchery. The method is followed also in recording the collection of eggs and planting of fry. Under this system, it is literally impossible to determine the loss of eggs and fry on any given lot, and the entire business savors of uncertainty and guesswork."[68]

Unable to reconcile his figures and those of the hatchery records, Ball reported the discrepancies and pointed out:

> ". . . assuming that these imaginary eggs would have hatched and that they with the imaginery fry . . . would have been liberated without loss, there would have accrued to the credit of Alaska Packers Association a tax exemption equivalent to $4,156.79."[69]

Table 10. Sockeye Production, 1896-1916: Karluk Hatchery

Brood Year	Spawning Period	Number of Eggs Taken	Number of Fry Released	Comments
1896	Aug. 29-Nov. 28	3,236,000	2,556,440	Spawning period are dates actual spawn taken.
1897	Aug. 19-Dec. 9	8,454,000	6,340,000	1% spawned in Aug.;85% in Sept. and 14% in Oct-Dec. Between Sept. 6-20, 65% spawned.
1898	June 1-Sept.	4,491,000	3,369,000	Spawning period are dates fish were caught.
1899	June-Sept.	10,496,900	7,872,000	
1900	June-Aug.	19,334,000	15,566,800	First free swimming fry released into Karluk River in January 1901.
1901	June 20-Oct.	32,900,000	28,700,000	Large run in Karluk River. Some eyed eggs and sac-fry placed in ponds during fall.
1902	June-Nov. 4	23,400,000	17,555,000	...
1903	June 11-Aug. 28	28,113,000	22,000,000	...
1904	...	45,500,000	33,670,000	...
1905	...	36,933,000	28,236,412	...
1906	July 2-Sept. 15	38,679,200	36,846,000	New superintendent in May. NMFS records show 33,844,000 fry released.[a]
1907	June 20-Sept. 20	47,808,200	43,655,000	NMFS: 37,250,000 fry released.
1908	...	40,320,000	36,075,000	NMFS: 30,700,000 fry released.
1909	June 30-Oct. 13	45,228,000	40,620,000	NMFS: 30,500,000 fry released.
1910	July 1-Sept. 15	49,626,000	37,722,000	NMFS: 31,150,000 fry released.
1911	July 1-Oct. 11	41,026,800	37,495,100	NMFS: 34,495,000 fry released.
1912	June 24-Sept. 20	45,500,000	41,803,155	New superintendent. NMFS records on fry releases agree with official numbers.
1913	June 21-Oct. 11	34,629,160	31,546,000	Some fry released into the Karluk River.

Table 10 — Continued

Brood Year	Spawning Period	Number of Eggs Taken	Number of Fry Released	Comments
1914	June 27-Sept. 30	30,240,000	27,704,000	Some fry released into the Karluk River.
1915	June 26-Sept. 29	41,135,000	23,948,000	Some fry released into the Karluk River.
1916	June	1,016,000	0	Station closed and eggs transferred to Afognak hatchery.

(a) NMFS — records cited refer to "Karluk Hatchery, 1903-1954," Auke Bay Fisheries Laboratory records. National Marine Fisheries Services, Federal Record Center, Seattle.

This was the first official disagreement between hatchery statistics and those submitted to the Bureau. But an envelope of hatchery data, which became part of the National Marine Fisheries Service records, includes a list of fry releases which do not agree with published data.[70] Both sets of data are shown in the table of production for the Karluk hatchery (Table 10).

Inconsistency in numbers was not the only problem Ball encountered. An epidemic of disease was in progress, but Ball did not try to assess the loss. Although the Bureau of Fisheries' annual report gives the loss at 41.8 percent, the actual percentage of fish lost is difficult to ascertain because of inaccurate record keeping, as Ball pointed out.[71]

Ball's report was sent to the Alaska Packers Association by the Bureau of Fisheries. No immediate answer was forthcoming so a telegram was sent by the Bureau informing the company of the discrepancies and asking for an explanation. Communications were slow and, during this time, Ball was instructed to again visit the hatchery. He did so in July. None of the hatchery personnel nor the superintendent of APA in Karluk were available. "It is natural," Ball wrote, "to conclude that he hurried to Kodiak to make oath to the hatchery returns before any further inspection could be made."[72] Ball reported the oath had been sworn before the U.S. Commissioner at Kodiak, and it showed nearly 10,000,000 more fry than Ball's estimates.

In the meantime at the executive level, APA responded to the Bureau's letters in late June, stating that "APA has always conducted its affairs upon such a high level of integrity and honesty that we are sure that the Bureau doesn't believe we are taking advantage."[73] The company sent Jefferson Moser to Washington, D.C. to go over the situation with Bureau officials. Moser told the Bureau that APA would accept anything that satisfied the Bureau, as Moser felt it would be difficult to resolve the numbers controversy because of the death of Superintendent Sargent.

Figure 22. Catch of Sockeye Salmon at Karluk and immediate vicinity from beginning of canning industry in 1882 to 1927. (Source: *Gilbert and Rich, 1929.*)

Then, on June 27, 1916, Moser wrote to the Commissioner of Fisheries: "In view of some unfavorable conditions which have developed in the progress of work at the hatchery, it has been deemed advisable by the Association to close the same at the end of fiscal year, June 30, 1916, and after the fry on hand are liberated."[74] The immediate answer from the Commissioner of Fisheries expressed his approval of the course of action described.

All that remained was to resolve the matter of the rebate. The Bureau notified the Clerk of the Court to withhold the rebate, but found that APA had not applied for it. Moser wrote to the Commissioner that the company would accept Ball's numbers, and the rebate was given on the adjusted numbers.[75] All Bureau of Fisheries fry release data show Ball's adjusted numbers.

The controversial Karluk hatchery was closed. The eggs on hand, 1,016,000, were sent to the Afognak hatchery. But the issue was not quite settled. The APA wrote asking to reopen the hatchery on March 7, 1917. The Bureau's answer was:

> "The Bureau regrets that you have raised the question in view of the spirit of cooperation which has been between APA and the Bureau. The Bureau has been a target of unfavorable criticism by the Delegate in Congress from Alaska and other people for its long continued approval of the Karluk hatchery . . . . The Bureau has upon occasion defended its course of action in not recommending withdrawal of approval of the Karluk hatchery when it became the object of attack and unfavorable comment several years ago, but the stand taken by the Bureau at that time was not upon any too firm of a basis . . . ."[76]

The Bureau denied approval to reopen the hatchery unless the hatchery was moved to the lake or a plan was made to transport the fry to the lake. The Bureau of Fisheries itself did consider, in 1917, the feasibility of that agency establishing a hatchery on Karluk Lake. The cost of developing a transportation system to Karluk Lake proved prohibitive.

It is difficult to analyze the influence, or even negative effect, the hatchery had on the Karluk system. The only attempt to determine adult returns by marking fry was not statistically designed and the results of the marking were unsubstantiated. The yearly commercial catches of sockeye salmon were researched by scientists Gilbert and Rich and the fluctuations in the runs continued before and after the influence of the Karluk hatchery as is shown in Figure 22.[77] The downward trend continued.

Three years after the hatchery closed Ball made an observation on the strengths of the runs and summarized the effects of artificial propagation on the Karluk system:

> "Excepting 1916 and 1917 when the catch was 200 percent greater than in any of the seasons immediately preceding, there has been but little fluctuation in the run of red salmon in recent years . . . . There is virtually no change in the amount of fishing carried on at Karluk each year . . . . It will be interesting to follow observations at Karluk through the next few seasons until the first run of salmon after the year in which the hatchery closed. The effect of artificial propagation at Karluk was regarded at all times as wholly negative, the run surviving neither for lack of hatchery operations or by virtue

of it. The 80,000 red salmon which were annually corralled at the hatchery, half of which died before they were ready to spawn, can now go to Karluk Lake and preform their life's function in nature's way and under ideal conditions.''[78]

REDFISH BAY

The first Southeast canning company to voluntarily begin artificial propagation of sockeye salmon was the Baranof Packing Company. During the summer of 1896 this company built a small hatchery at the head of Tumakof Lake, which drains into Redfish Bay on the west coast of Baranof Island.

This bay, about 60 miles south of Sitka, was the site of the company's cannery, which first packed in 1891. It was a small cannery with no regular fishing stations. The company steamer went from stream to stream where it believed that fish could be obtained and, when loaded, returned to the cannery. The streams were scattered over a territory fished by no other canneries and ranged on the outer coast from Cape Ommaney to Cross Sound and on both sides of Chatham Strait from Icy Strait to Cape Ommaney. This was considered one of the hardest fishing routes in Alaska. The company took a large number of sockeye each year at the mouth of the home stream, the average catch during a seven year period being 34,903 sockeye. The largest catch prior to 1897 was 69,553 fish in 1894.[79]

There was sufficient concern about this stream's output that the lake and its tributaries were examined and the phenomenon of lake-spawning sockeye was discovered. In those early years it was assumed that all sockeye salmon spawned only in tributaries.[80] It was also found that coho salmon spawned on the shelves of the lake which subsequently dried up in winter.

Moser visited in 1897, the year after the construction of the hatchery, and found a dam on the lake outlet which had a controlled opening six feet wide. He was told by company personnel that the tributaries froze each year, and the lake level dropped. To prevent this the dam would be closed after all fish had gone up, maintaining the lake level high enough to keep the eggs covered. However, this also was the first year that stream barricades were illegal and it is unrecorded whether the plan was actually executed.

Company officials told Moser that they appreciated the value of the home stream and wanted to increase the output; therefore, a hatchery had been built during the summer of 1896 and finished in time for experimental work in the fall of that year. The hatchery was located at the head of the lake and consisted of a log and rough-board building, 20 feet by 30 feet (Fig. 23).

There were eight hatching troughs, each 16 feet long, 12 inches wide and 11 inches deep, made of unplaned and unpainted boards, one inch thick and 12

Figure 23. Redfish Bay hatchery, 1897. *U.S. Fish Commission.*

inches wide. The troughs were arranged in three pairs, one against the wall on either side of the building, leaving four passageways between them. There were six compartments in each trough, 30 inches long, separated by solid wooden partitions six inches high, joined tightly to the bottom, and the water was allowed to flow only over the top. In each compartment were two, one-eighth inch round iron rods, fixed transversely one-fourth inch from the bottom, for the baskets to rest upon. The baskets were 24 inches by 10.5 inches by 6 inches and held about 25,000 eggs.

Water from a falls, a few hundred yards behind the hatchery building, was connected by a box flume to the northeast corner of the building and flowed into a transverse distributing trough. It then passed through the troughs and out a wastepipe.

Operations began in the fall of 1896 and after 200,000 eggs had been taken, unexpected cold weather set in and not only froze the water in the flume solid, but froze the whole falls. The eggs were put into the lake and left to their fate. It was the intent in 1897, in the event of a freeze, to pump water from the lake and use a filter. A boiler and pump were installed. The stream water, when tested, was about eight degrees colder than that in the lake.

It is unrecorded whether eggs were taken in 1897. If so, it was the last time, as Baranof Packing Company sold its cannery to the Alaska Packers Association in 1898, and the buildings were subsequently dismantled. The new owners were not interested in operating the tiny hatchery on the remote, treacherous west coast of Baranof Island, so far from any of their active canneries.

KLAWAK HATCHERY

The management of another canning company in Southeast Alaska, the North Pacific Trading and Packing Company of Klawock, realized the value of their home stream and decided to improve the catch by starting a hatchery. It was to John C. Callbreath, at McHenry Inlet that the superintendent of the company looked for advice. Callbreath gave counsel on site selection and necessary hatching equipment. He advised the cannery superintendent ". . . to check the little streams on the lake and find one that is warm as that means spring water which is pure and relieves the expenses of building a filter and the temperature being even makes the eggs hatch better."[81] Callbreath loaned the company the services of his fish culturist to aid in site selection and thus it was Fred Patching who selected the first hatchery site at Klawak Lake.

In May and June, 1897 erection of a hatchery, with a dwelling for the attendant, began on the right bank of the outlet, close to Klawak Lake, immediately below the upper rapids (Map 6). This board and batten structure was 50 feet by 16 feet with a spruce shingle roof (Fig. 24). The hatchery fixtures consisted of 8 troughs, 16 feet long by 16 inches wide, placed in pairs so that there were two sets of troughs running 32 feet in length (Fig. 25). These were separated in the middle, the two lower pairs in each set being four inches lower, allowing that much fall to the water for aeration. (Fig. 26).

The water was conducted from the lake in a V-shaped trough to the upper end of the building. After passing through a large sluice box, it ran into a head box above the hatching troughs. Waste water returned to the stream through iron pipes at the foot of the troughs.

These troughs were of the regular Williamson design, of planed lumber, well coated with asphaltum varnish. The troughs had compartments for 56 baskets, with each basket designed to hold 50,000 eggs, giving a trough capacity of 2,800,000 eggs.

The hatchery operated in the fall of 1897 and of the 2,023,000 sockeye eggs taken, only 800,000 fry hatched. The water supply was the major problem. During the first part of the incubation period, the surface water of the lake was 65 °F. Eggs eyed and hatched quickly, but soon died.[82] A fungus and/or slime began to gather on the wire of the baskets and caused mortality when the eggs stuck together. A make-shift filter was installed in an attempt to alleviate this. This filter consisted of a layer of coarse gravel and charcoal and then layer upon layer of finer material. When Moser visited in 1897, he thought the filtering system only partially corrected the problem because it was badly arranged. Water entered along the bottom and rose about half-way up the filter and then

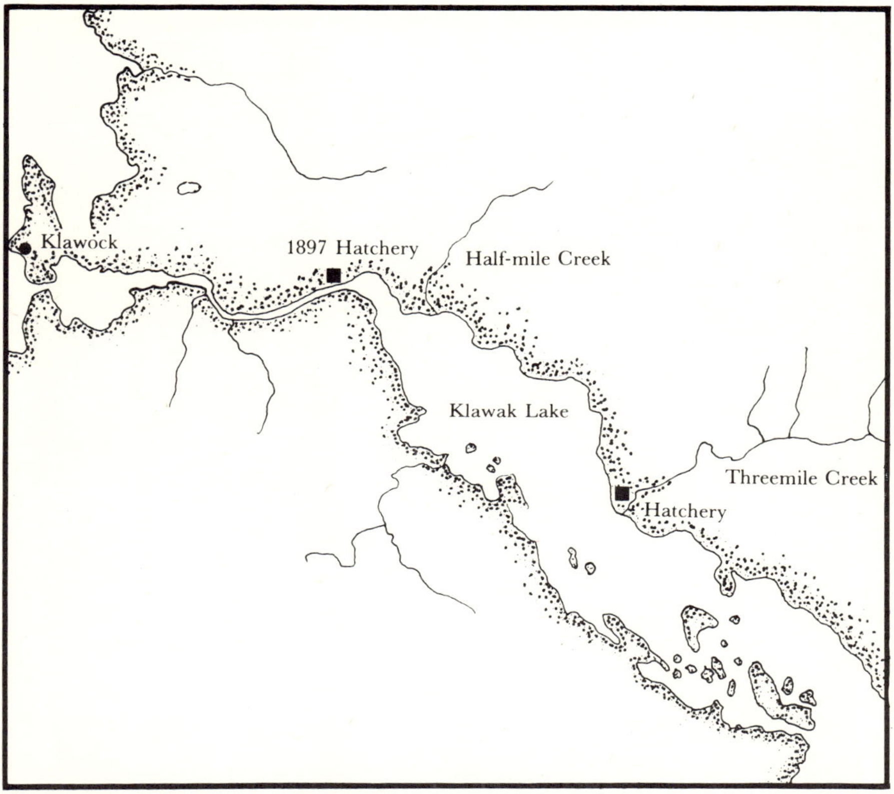

Map 6. Klawak Lake and Hatchery Locations

was delivered. Thus, the water did not reach the finer material. He suggested a box or tank filter of sufficient height to let the water enter by a spreader and percolate through the entire course. He also thought the flow of water seemed small and aeration of water insufficient.[83]

The hatchery met the standards of the time and Moser described his impressions as:

> "The manager has spared nothing in perfecting the plant and making the hatchery a success. The superintendent is a man who was employed by Callbreath and methods are those used at McHenry Inlet. The fish are stripped according to methods used at Battle Creek [California].
>
> "About a pint of water is put in a pan, the female stripped into this pan and the male directly over it. Eggs are not picked over immediately."[84]

Figure 24. Original Klawak hatchery on Klawak River, 1897. Water conveyed from lake by flume. *National Archives.*

Figure 25. Interior of original Klawak hatchery showing covered troughs, 1897. *National Archives.*

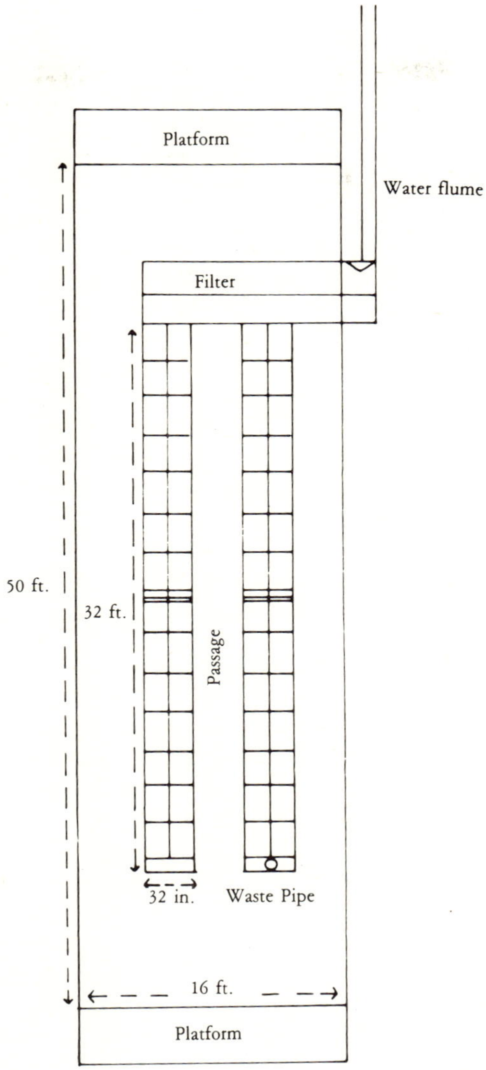

Figure 26. Plan for the original 1897 Klawak Hatchery, located near the outlet of Klawak Lake; dismantled after one year's operation.

Most of the fish were obtained at Half-Mile and Threemile creeks on the northern side of Klawak Lake. Fish were captured in a stake trap, removed by dip net and, if ripe, spawned. If not, they were returned to the lake. Fertilized eggs were carried by boat to the hatchery.

After the first year's experience, it was decided to move the hatchery to Threemile Creek, three miles east of the former location, a distance of six miles from the North Pacific Trading and Packing Company's cannery at Klawock. The original hatchery was dismantled for lumber, and a new hatchery 60 feet by 18 feet was erected. It contained 12 troughs, 16 feet by 14 inches arranged in pairs. There were six baskets to a trough, yielding a capacity of 3,600,000 eggs. The capacity was further enlarged when a new hatchery was erected in 1908 (Fig. 27). Two years later, in 1910, 20 troughs were set crosswise in the building, giving a total capacity of eight million eggs with the possibility of expansion if necessary by addition of troughs.

The hatchery water was supplied by a 1,800 foot long wooden flume from the head of a feeder stream of Threemile Creek. This was replaced in 1902 after a strong wind storm downed trees across the flume. Four and one-half inch pipe was used for 600 feet to connect with 347 feet of flume. A three-foot dam created a small pool and the water flowed through a small gate which controlled the quantity, thus reducing the head.

A picket fence was built across Threemile Creek near the mouth to prevent salmon from going upstream and to facilitate capture of adult salmon. A straight, coarse-meshed seine was used to capture the fish. Only the ripe females and a sufficient number of males were kept. To give the returning adult salmon a better chance to migrate into the lake, the canning company in 1915 blasted out a number of rocks which partially obstructed the outlet of Klawak Lake. It was expected this would also keep the lake from rising during storms to such a height that it allowed fish to pass the racks at the tributaries. It may be coincidental that the following year the largest number of eggs in the hatchery's existence was taken.

The fish were generally spawned by the pressure method, the incision method not being used until 1916, the last year of operations. Spawned fish were dumped into the middle of the lake. The eggs were fertilized immediately with milt from 2.5 to 3 males per female. Water was added within a few minutes after fertilization, and the pan was set for 1.5 hours before washing and placement of eggs in the troughs. Egg enumeration was based upon a reckoning of 3,500 eggs per female until 1911 when exact measurements began. Eggs were picked when first put into the hatchery and were not disturbed for 10 to 15 days. At that time picking resumed. When egg spots appeared, eggs were feathered each day to prevent suffocation.

Cannery management told Fisheries Agent Coyle in 1902 that the water in the troughs was artificially heated from a steam coil which ran through the stove and then through the hatchery troughs. He was also informed that the incubation period averaged 100 days.[85] These numbers disagree with those given to another federal employee, William Hancock, superintendent of the Yes Bay hatchery. He was told the first eggs in the hatchery hatched in 77 days, or 1266

Figure 27. Klawak hatchery at Threemile Creek, 1912. *National Archives.*

T.U. at 52 °F maximum and 42 °F minimum. There is no mention of heating trough water.

Difficulties with freezing of the water system undoubtedly prompted the management to discuss heating water. The water system froze each winter from 1900 through 1902. The pipeline was laid on top of the ground and the cribbing around it filled with dirt and moss, but this was not sufficient to prevent freezing when cold weather occurred. In all cases the eggs and/or sac-fry were planted immediately in the lake, although it is unclear whether the eggs taken in 1900 were frozen.[86]

No fry were held until the free-swimming stage nor was there an attempt to feed fry during the first sixteen years. Sac-fry were held in the troughs, usually for about two weeks, but never until the yolk-sac was absorbed. If the eggs hatched in large quantities, it was necessary to plant the sac-fry in less time as they tended to pile up and smother.

The methods for releasing the sac-fry were described by company officials to Fisheries Agent Fassett during inspection prior to certification under the 1906 act. He was told that releases were made around the margins of the lake and shores of islets in depths of three to five feet, where a clean rocky bottom gave good cover.[87] Mouths of feeder streams were avoided because of possible presence of trout, sculpins and other predators. Places where aquatic birds congregated were also avoided. Fassett was told that a search was always made in advance to determine the presence of predators and these places were avoided.

When sac-fry were to be released, 20,000 to 25,000 sac-fry were measured into a galvanized iron bucket coated with asphaltum lacquer or varnish. Crowding was carefully avoided, Fassett was assured. To liberate the sac-fry, one man rowed the boat as slowly as possible while the superintendent sat in the stern and held the bucket of sac-fry down into the water, with the after edge just lipping the surface of the lake. The sac-fry wiggled over the lip of the bucket a few at a time and sounded at once. Occasionally, if very close to the shore, they made for land, but this was not usual.

Once a release was made in a particular place, no more sac-fry were liberated in that vicinity again for a period of five to six weeks. Hatchery personnel believed that much time was required for the yolk-sac to absorb and once this had taken place, the fry made off into deeper water, leaving the area free once more.

The method of releasing the sac-fry was not approved by everyone in the Bureau of Fisheries. After the 1906 act went into effect, requiring certification of efficiency, Claudius Wallich, field superintendent of the federal hatchery at Yes Bay, was asked to make an inspection.

Wallich wished to postpone his visit until spring when sac-fry were being liberated so that the report would reflect that the hatchery habitually planted sac-fry as soon as hatched with no attempt to hold fish to the extent of hatchery capacity. Wallich wrote the Commissioner of Fisheries: "It is not good to plant fry as soon as hatched and it is respectfully suggested that it should be a condition precedent to a favorable report, that at least a portion of the fry be held until the final absorption of the sac."[88] To stress his point, he went on to say:

> "The North Pacific Trading and Packing Co. was informed, first informally by me, and later through the inspection directions of the Bureau, in regard to some of the requirements, yet they never indicated in any manner that they have in the past, or intend to in the future, be guided by our desires. It is easily possible and is perhaps more frequently done than is generally known, by means of indifferent hatchery methods, to hatch and plant the entire product of a season's take of which perhaps not 5 percent will reach the free swimming stage."[89]

Subsequently Wallich sent North Pacific Trading and Packing Company a letter stating requirements for certification. These requirements were that as large a number of sac-fry as possible had to be held until the sac was absorbed, and a heating plant was to be installed.[90] But Wallich's directives were not followed after he was transferred to Afognak Island to supervise construction of that hatchery in 1907. His replacement, William Hancock, the new Yes Bay superintendent, completed the inspection for certification of efficiency. Hancock made his tour when there were no eggs or sac-fry in the hatchery, despite Wallich's recommendation. Hancock found essentially the same thing which Fassett reported in his letter. The fact that the sac-fry were liberated between October 26 and January 4, 1907 was not stressed. Hancock's only comment was "the shores were lined with cobble stones making excellent hiding places."[91]

The hatchery continued to plant sac-fry for another seven years, until 1913. In that summer, a pond 16 feet by 24 feet was constructed and first used on the 1913 brood. It was too small to accommodate large numbers of fish and sometimes froze to the bottom. The majority of the sac-fry continued to be planted along the lake shores and, in 1915 and 1916, in the small streams tributary to the lake.

It appears that, with frequent changes in superintendents, much of the original techniques for releasing fish were ignored. A fisheries agent, during a visit in 1915, devoted special attention to the conditions under which sac-fry were liberated:

> "They [the attendants] were made to realize that hatchery efficiency does not consist merely on liberating a stated number of young fish, but rather that the percentage of those surviving until they are finally able to take care of themselves is the real basis for determining the efficiency of all fish cultural work. Unusual interest was shown by the hatchery employees in the selection of suitable locations for the planting of young salmon and the exercise of judgment as to the time of planting, so that the greatest possible number of fish might survive."[92]

These technical inefficiencies, however, were not what closed the hatchery. Rather it was the inability to obtain and keep a hatchery superintendent. Superintendents changed almost yearly. The hatchery superintendent in 1917 left at the last minute and no replacement could be found in time for the spawning season. In 1919 the North Pacific Trading and Packing Company, when queried on their plans for the hatchery, told the Bureau that it was their intention to operate, but it was dependent upon ability to find the right party to put in charge. Since the hatchery did not reopen, it is assumed that no such person was found.

Only one attempt was made during the years the hatchery operated to determine the success of the program. Fry were marked by fin clipping in 1899 and results of this study have been previously discussed.

The impact of the Klawak hatchery on the commercial fishery appears to be negligible. Commercial catch data for Klawak Inlet are shown in Figure 28. The number of eggs taken was small and heavy losses occurred at times from freezing. Tables 11 and 12 show the number of eggs taken and fry released.

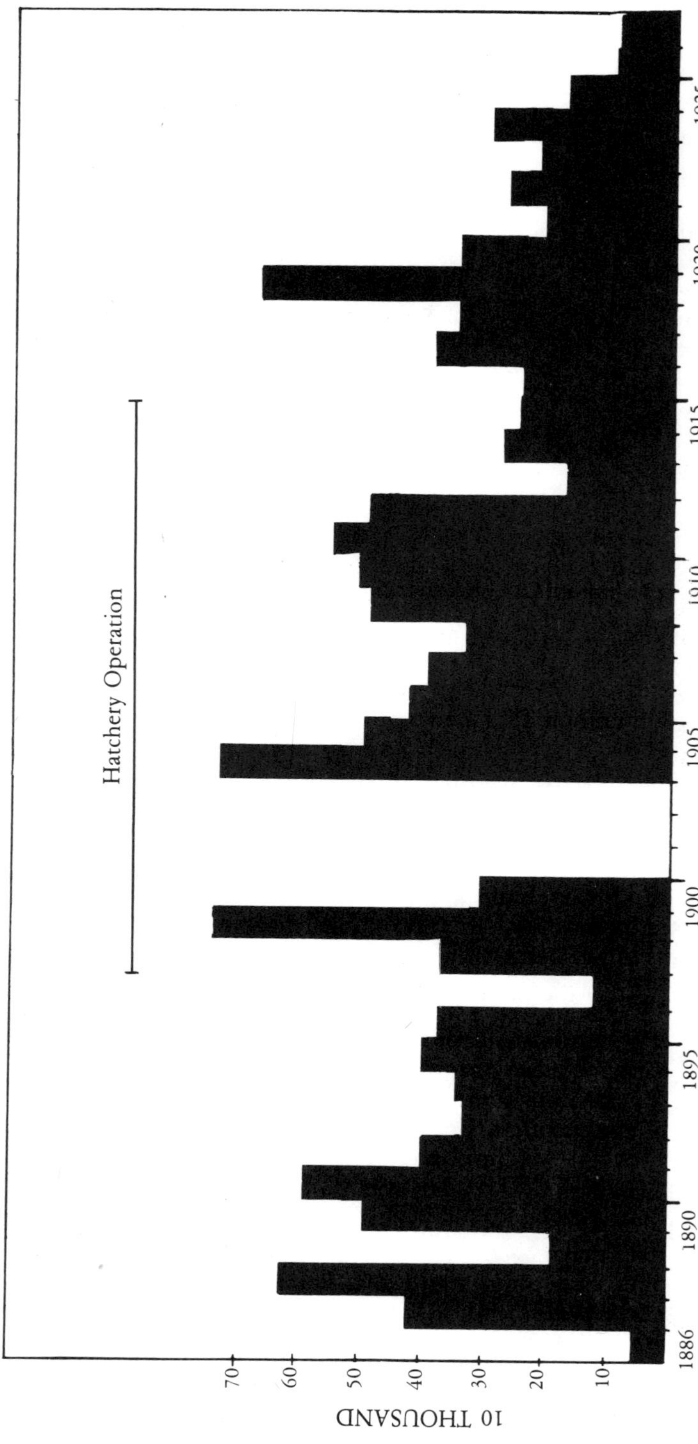

Figure 28. Catch of Sockeye Salmon in Klawak Inlet, 1886-1900, 1904-1927. No data available for 1901-1903. (Source: *Rich and Ball, 1935.*)

Table 11. Sockeye Spawn Taken, 1897-1916: Klawak Hatchery

Brood Year	Number of Eggs Taken	Spawning Period	Limiting Factors
1897	2,023,000	Aug. 1-Sept. 14	Weir washed out.
1898	2,586,000	July 23-Aug. 26	Hatchery capacity.
1899	3,538,000	...	Hatchery capacity
1900	3,600,000	...	Hatchery capacity
1901	3,500,000	...	Hatchery capacity
1902	1,470,000	Sept. 5-Sept. 26	High water prevented placement of weir, then weir washed out.
1903	3,000,000	...	...
1904	2,800,000	...	Heavy storms washed out weir and dams.
1905	2,800,000	...	Drought caused run to be late. Then high water washed out weir.
1906	3,360,000	Aug. 5-unknown	Spawned 15% of females entering lake and 40% in Threemile Creek.
1907	3,500,000	...	Hatchery capacity. Used 960 females.
1908	3,500,000	...	...
1909	...	...	No report to Bureau of Fisheries.
1910	6,786,500	Aug. 6-Oct. 17	Lack of spawners.
1911	5,600,000	...	Lack of spawners
1912	3,835,000	Aug. 16-Sept. 29	Lack of spawners.
1913	3,645,000	Aug. 8-Sept. 28	Lack of spawners.
1914	3,816,000	Aug. 6-Oct. 2	Lack of spawners.
1915	4,130,000	July 24-Nov. 9	Lack of spawners.
1916	8,160,000	July 20-Sept. 26	Hatchery capacity.

Although the Klawak hatchery's contribution to sockeye runs remains doubtful, the selection by the early cannery managers and fish culturists of Klawak Lake as an ideal hatchery site is more solid. Bureau of Fisheries scientists, such as Moser in 1897, made such statements as "If the government should ever desire to establish a hatchery in this region, it is believed that here the greatest success could be obtained with the least expenditure of money."[93] In 1903 Wisner said "I believe that it will be difficult to locate a better place in the vicinity than Klawak for a hatchery."[94] The Alaska Department of Fish and Game came to the same conclusion, and a state-operated chum and coho salmon hatchery operated for the first time in 1978. Its location is within 100 feet of the original 1897 North Pacific Trading and Packing Company building.

Table 12. Sockeye Released, 1897-1916: Klawak Hatchery

Brood Year	Fish Released	Dates of Release	Distribution
1897	800,000	Oct. 12-unknown	Planted as sac-fry.
1898	2,150,000	...	Planted as sac-fry in shallow parts of the lake.
1899	2,180,000	...	Planted as sac-fry in shallow parts of the lake.
1900	0	0	Severe windstorm broke water flume. All stock planted immediately. Some sources say all eggs frozen.
1901	1,113,000	...	Severe winter. Ice in hatchery. Questionable if any release made.
1902	1,155,000	...	Water system froze, eggs and sac-fry planted immediately.
1903	1,700,000	...	Unknown distribution.
1904	2,000,000	...	Unknown distribution.
1905	2,000,000	...	Flume carried away in flood. Eggs planted in lake.
1906	1,187,000	Oct. 26-Dec. 27 Jan. 4	Planted along shore as sac-fry Water system froze and eggs about-to-hatch were planted.
1907	2,776,000	...	Unknown distribution
1908	3,200,000	...	Unknown distribution.
1909	5,300,000	...	Unknown distribution.
1910	6,200,000	Nov.-April	Unknown distribution.
1911	3,530,000	Dec.-June 30	Unknown distribution.
1912	3,675,000	Nov. 28-Feb. 4	Unknown Distribution.
1913	3,465,000	Jan. 1-May 1	Sac-fry placed in rearing ponds for first time.
1914	3,653,000	Dec. 14-April 15	Some sac-fry rearing in ponds.
1915	4,020,000	Oct. 6-May 7	Planted in small tributaries to Klawak Lake. Some sac-fry retained in ponds.
1916	7,822,000	Sept. 26-Feb. 16	Some sac-fry retained in ponds.

HETTA HATCHERY

Another facility established before hatcheries became mandatory was built by the Pacific Steam Whaling Company at Hetta Lake. The mouth of this stream was historically a favorite spot of sockeye fishermen and was considered one of the best sockeye streams in Southeast Alaska prior to the turn of the century. This system was a natural location for construction of a hatchery by a canning company interested in maintaining a lucrative run of these salmon.

The Pacific Steam Whaling Company, which operated a cannery at nearby Hunter Bay, voluntarily began erection of a hatchery building on a site on the stream between Hetta Lake and Hetta Cove in 1898. Work continued through the season but the building was only partially completed. When Fisheries Agent Willliam S. Ballard arrived on the 1898 yearly inspection, he found that the company had begun to construct a series of fenced ponds within the main stream to collect spawners. The first fence was located at the upper end of the hatchery building, and was made of wire net and stakes. The second, 100 feet below the first, was made of stone and formed a stone wall across the stream except for about six feet in the center where a gate was constructed. The third, about the same distance from the second, was built in the same manner with the gate six feet on one side of the stream.

Pacific Steam Whaling Company had sought and obtained permission to fence the stream at the point near the hatchery two years after an act had made stream barricades illegal. This exception to the law specified that the fencing was to be so placed as not to impair the rights of other parties interested in fishing at Hetta and that no salmon so impounded were to be used in the cannery. But since the company had fenced the stream so tightly that it would be impossible for the smallest fish to get into the lake, thus impairing the rights of other parties, the company was instructed to remove the barricades.

The following year, 1899, the hatchery was moved, possibly because a different method for holding adults until ripe needed to be developed. It had also been found that the water could not be taken from the lake as planned. The new site was located on low, rolling, and heavily wooded land on the southeast shore of the lake, on the left bank of a feeder stream which does not show on today's topographic map (Map 7).

Operations at Hetta can be divided into two periods: 1899 to 1903 and 1908 to 1918. During the first period the hatchery was not operated in a wholehearted manner. Records are scanty and Wisner did not believe a serious attempt was made to secure all the eggs possible or to secure the best labor possible for handling them.[95] During this period the hatchery consisted of a plank building 50 feet by 26 feet. There were 10 Williamson troughs, each 20 feet long, 12 inches wide and 8 inches deep. The basket divisions were two and one-half feet in length and two division plates, four inches apart, were made of asphalted light-weight iron. The headbox was 12 inches wide and of the same

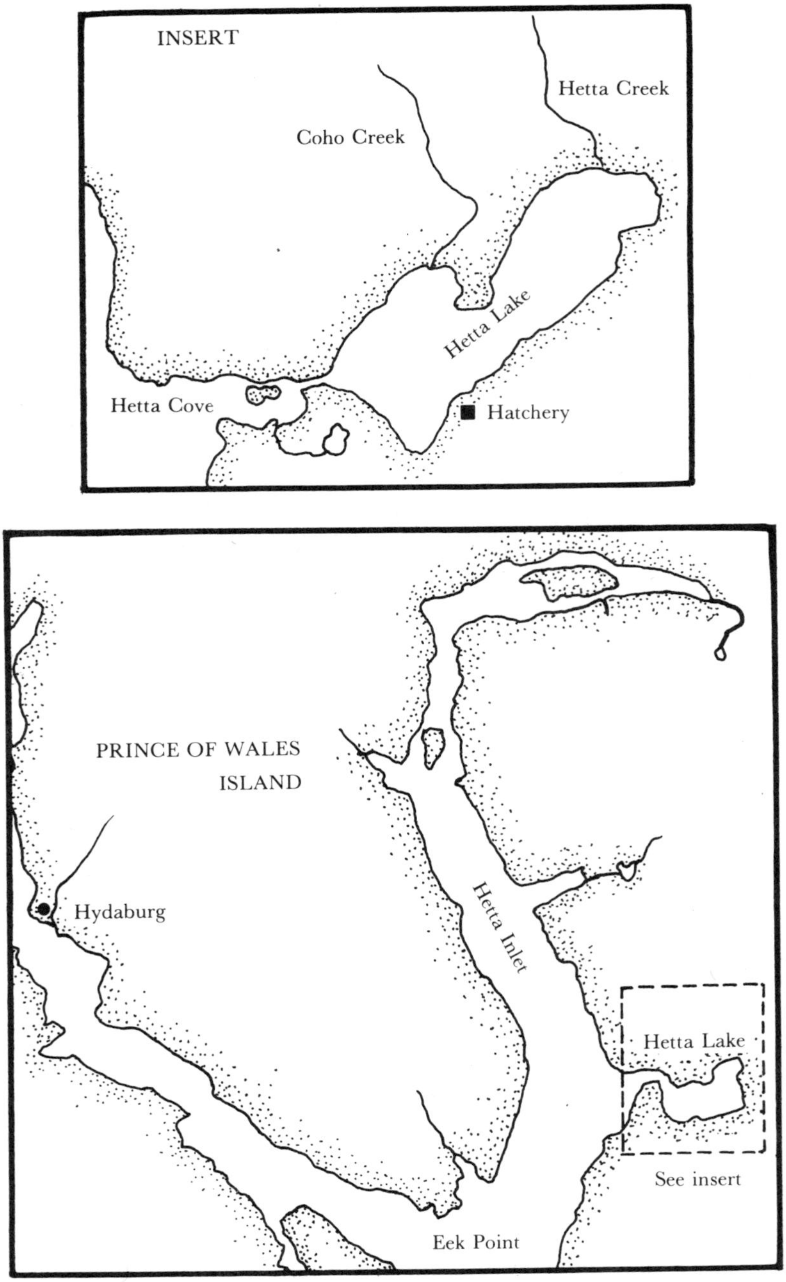

Map 7. Hetta Inlet and Hetta Cove and Lake

depth. Gates were made by cutting out and resetting the block, leaving a space at the bottom of three-fourths inch by four inches. A sheet-iron drop gate controlled the outflow through a short in-spout into the troughs, a fall of four inches. In the double row of troughs a similar spout carried the water into the second section with a fall of two inches. The troughs dropped about six inches in their length.

The baskets were 24 inches by 11 inches by 5 inches, of the usual material, two-tenths inch by three-fourths inch mesh. Wooden battens were secured to the two long sides, the ends protected by a fold of tin nailed to the battens, and the corners interlaced. The baskets were supported in the troughs either by the wooden battens or upon four small blocks one-half inch high, nailed to the bottom of the troughs. The 70 baskets in the hatchery each had a capacity of 50,000 eggs.

The water supply came from a spring-fed stream by the hatchery. A flume 325 feet long carried the water across country from a ten foot cascade about 200 yards above the stream mouth to a settling pond. The 12 foot drop was not sufficient to keep ice from forming during periods of freezing weather. On one occasion ice gathered to a height of two feet, and a notch had to be chipped out three times in a 24 hour period to keep a small flow in the hatchery.

Difficulties with freezing water in 1900-1901 accounted for a large loss of eggs. The records for that season are incomplete and inconsistent, and the output is uncertain. The next season, 1901-1902, the water again froze, and the sac-fry and remaining unhatched eggs were planted in March. A large stove, made from a coal oil tank, was used to heat the building in an attempt to keep the water in the troughs from freezing. When this proved unsuccessful, a second stove was installed. Wisner was told that a larger stove with a coil to the headbox was to be added, but this apparently was not done.[96]

The spawners during the first phase of the operations were seined and spawned by the wet method. Eggs were measured by use of a bucket which held 52,500 eggs, but fecundity was calculated on an average of three females, with numbers rounded to 3,500 per female. Wisner learned in 1903 that the period of incubation for eggs taken August 20, 1902 was 128 days, but eggs taken in October took between 191 and 210 days. In some years sac-fry were being released at the same time that the next year's adults returned to the lake.

Those fisheries officials who visited the hatchery during the period from 1899 to 1903 were not satisfied with the fish culture methods in use. A.J. Young, in charge during most of this period, appears to have had little supervision and minimal help. Fisheries Agent Coyle learned in 1903 that baskets of eggs were not covered during incubation because Young felt that it was unnecessary. Young also believed ". . . that the best time for planting fry was in March and that fry should be kept in the hatchery for four weeks or until the sac is absorbed."[97] Young told Coyle that fish were planted at one and one-half inches in length after being fed liver boiled hard and chopped fine and sometimes mixed with wheat bran.

Young may have known the correct techniques, but surviving records show that in 1899 the sac-fry were placed in the stream prior to January 22, 1900 and that in 1900 and 1901, eggs were either frozen or were planted as eyed eggs or sac-fry. It is unlikely that significant numbers of sac-fry, if any, were held during this period despite what Young told Coyle.

The hatchery closed for the four years from 1904 through 1907. This closure did not reflect on its inadequacy, but resulted from changes in ownership of the cannery at Hunter Bay. Pacific Steam Whaling Company joined the Pacific Packing and Navigation Company, which eventually declared bankruptcy. When Northwestern Fisheries Company acquired the interests, the Hetta hatchery became part of that company's assets.

After being idle for four years in the moist climate of Southeast Alaska, the hatchery building and its equipment needed renovation and in the process improvements were completed. The water supply was protected against freezing by placing a wooden four inch pipe two feet underground from the point of intake in the creek to the hatchery. The hatchery capacity was enlarged to ten million eggs. Certification necessary to qualify for continued operation and for rebates was granted in January 1908, after a very cursory examination by a fisheries agent on October 25, 1907, a year before the hatchery began to incubate eggs again.

Operations were again placed under the supervision of A.J. Young until the fall of 1911 when he was replaced by Thorgrim Casperson, a long-time employee at the Yes Bay hatchery. Young may have again run the hatchery some years before it closed.

Unlike other locations where hatcheries operated, Hetta Lake had no major tributary where a weir could be erected, so adult salmon capture proved difficult. The logbook from the hatchery shows that the sockeye salmon began to run in June, but it was August and September before the fish began to appear at the tributaries across the lake from the hatchery. Superintendent Young observed that the fish then would move from this portion of the lake closer to the hatchery. Seepages and springs entering the lake kept the water from freezing and sockeye salmon spawned along the shores. Fish would be observed and captured in small numbers up until mid-December. Young saw sockeye spawning in March.[98]

Securing spawning salmon which were scattered throughout the lake was a difficult and tedious job. Most of the superintendents did what they considered their best in order to secure salmon spawn at a time of the year when storms, downpours, or freezing and snowy weather were normal. It also took time and effort to clear places where beach seines could be used at the mouths of tributaries.

Table 13. Sockeye Spawn Taken, 1899-1917: Hetta Hatchery

Brood Year	Number of Eggs Taken	Spawning Period	Comments
1899	2,800,000	...	...
1900	1,800,000	Sept. 12-unknown	High water prevented adult capture.
1901	1,608,000	Aug. 20-Dec.	859 females spawned.
1902	4,088,000	...	...
1903	5,127,500	Aug. 17-unknown	...
1904	0	0	Hatchery not operated.
1905	0	0	Hatchery not operated.
1906	0	0	Hatchery not operated.
1907	0	0	Hatchery not operated.
1908	8,400,000		Hatchery capacity enlarged.
1909	10,313,000	Aug. 16-Dec. 19	...
1910	9,141,000	Aug. 16-Dec. 18	...
1911	2,575,000	Sept. 3-Nov. 16	Late start and then early lake ice limited fish captured.
1912	3,780,000	Sept. 1-Dec. 14	New hatchery with larger capacity.
1913	4,112,000	Aug. 8-Jan. 7,1914	...
1914	7,438,500	Aug. 10-Dec. 18	First eggs taken hatched Nov. 7 in 94 days.
1915	8,114,000	Aug. 19-Dec. 24	First eggs taken hatched Nov. 19 in 94 days. Eggs taken Oct. 4 & 7, hatched March 21.
1916	3,247,000	Aug. 21-Dec. 20	...
1917	4,826,000	Aug. 15-Dec. 14	...

If Young had responded to the accusation that he did not try hard enough to secure spawning salmon, he probably would have stated that such extensive operations required more manpower. Records reveal that two men were the usual staff, one of whom was engaged practically throughout the year (the superintendent) and one or two others who were required for several months. [99] Fisheries Agent Cobb reported in 1910 that Young's wife and six children did much of the work, receiving $2 a working day.[100]

Once fish were caught, they were immediately spawned and eggs placed in hatchery baskets. The number of eggs taken annually is shown in Table 13. Incubation procedures were standard for the time and some incubation data are available in the logbook but are not summarized.[101]

Table 14. Sockeye Released, 1899-1917: Hetta Hatchery

Brood Year	Fish Released	Dates of Release	Distribution
1899	2,600,000	Unknown-Jan. 22	Placed in streams as sac fry.
1900	...	...	Many eggs froze and some eggs put in stream to hatch. Output uncertain.
1901	...	March	Sac-fry planted because of freezing weather.
1902	...	...	No two sources agree on number of sac-fry released.
1903	...	Unknown-June 30	Last eggs to hatch released as sac-fry.
1904	0	0	Hatchery not operated.
1905	0	0	Hatchery not operated.
1906	0	0	Hatchery not operated.
1907	0	0	Hatchery not operated.
1908	8,143,000		
1909	9,000,000	Feb. 12-July 31	...
1910	8,552,500	Dec. 27-July	...
1911	2,342,000	April 10-18	...
1912	3,592,000	March 1-July 16	...
1913	3,590,000	Feb 22	...
		March 24	1,800,000 sac-fry released in pond.
		April 22	...
		April 23	Released in south end of lake.
		April 24	Released in northwest end of lake.
		April 25-June 15	Released 345,000 in lake.
1914	7,142,500	Feb.	Total release includes 23,000 from 1913 brood.
		March 3-April 12	3,000,000 released in ponds.
		April 3	October spawn — 500,000 fry released head of lake.
		April 15	October spawn — 1 million fry released head of lake.
		April 17	Released 700,000 at unnamed creek across lake from hatchery from Nov. spawn.
		April 18	Released 300,000 east side of lake.
		May 13	617,000 released on west side of lake.[a]
1915	7,598,000	Mar. 11	Released 1 million in hatchery stream.
		April 30	Released 50,000 along shores at foot of lake.
		May 18-June 17	Released 4,100,000.
		June 4	Released 1 million east side of lake.[a]

Table 14 — Continued

Brood Year	Fish Released	Dates of Release	Distribution
1916	3,120,000	Mar. 7 & May 29	Released 1 million total at unnamed creek.
		May 18	Released 1 million in Hetta Creek at head of lake.
		June 15-22	Released 620,000 in rearing ponds.[a]
1917	4,587,000	Feb. 7-Mar. 29	Released 3,250,000 at foot of lake and in hatchery stream.
		Apr. 29-May 27	Released 1,337,000 in rearing ponds.[a]

[a] Individual releases do not add up to total release in original Hetta Hatchery logbook.

During Young's second superintendency, Cobb visited Hetta. He reported that sac-fry passed out of the building by means of a trough into an oblong box set into the ground. From there sac-fry moved into the small creek only a few yards from the lake. A wire corral was built in the lake around the mouth of the creek and all sac-fry were held there until the sac was absorbed. [102] A soft gillnet placed over the corral protected the sac-fry from predators. This system was used only a few seasons. Feeding "do-overs" from the cannery was attempted with indifferent success.

After Casperson was placed in charge, it was reported that

> ". . . somewhat old-fashioned methods have prevailed, but beginning this year [1911] a change was inaugurated whereby improvements, including taking [eggs] by incision will be adopted. Also the inaccurate and undesireable plan of basing the count of eggs upon the supposed average of 3,500 for each female stripped will be discontinued in favor of actual measurement by dipper."[103]

The planting of fry, it was also reported, would be, henceforth, in other suitable parts of the lake. Up until that time all sac-fry were released at the hatchery stream. Table 14 shows the number of fry released.

A major reconstruction was undertaken in 1912. A new building was built on approximately the same site, utilizing such lumber as could be salvaged from the old hatchery (Fig. 29). The old troughs, which were still usable, were installed and new troughs added to bring capacity to 14 million eggs (Fig. 30). A new four inch pipeline, 310 feet long, was installed. The storage dam was repaired and height added to increase its capacity and a pond was constructed to hold sac-fry until ready for release.

Figure 29. Hetta Hatchery Buildings, 1910. *Cobb collection, University of Washington.*

Figure 30. Interior of Hetta Hatchery, 1914. *Bureau of Fisheries.*

The closure of the hatchery occurred at the end of the 1917 brood release in 1918. No announcement was made, and if the Bureau of Fisheries was notified, correspondence has not survived. There is, however, sufficient correspondence prior to the closure to speculate why Northwestern Fisheries Company chose to discontinue hatchery operations at Hetta despite the fact they operated their Quadra hatchery for many years. One factor was undoubtedly the Bureau of Fisheries officials' insistence that hatchery operations be conducted with newly developed techniques.

Two inspection trips were made of the Hetta hatchery by Fisheries Agent E.P. Walker in 1916. On the first inspection Walker found sediment in suspension in the water system. Repeated suggestions to hatchery and company personnel on how to solve the problem met with little success. One nursery pond had been installed by 1916, but Walker found it inadequate. The pond leaked water and fry and it was largely above ground where it accumulated snow and ice. Sac-fry continued to be liberated directly into the lake margins.

On his second visit, Walker found baskets of eggs had not been regularly picked, and eggs were bunched with fungus. In the bottom of several troughs where sac-fry were being held, large fungal-covered masses of dead eggs and fry were found, and in some cases this mass covered the entire bottom of the trough an inch deep.

The Commissioner of Fisheries wrote to Northwestern Fisheries Company in the spring of 1917 to suggest installation of both a settling pond and a small dam around the intake to reduce sediment in the water. It was also suggested a series of small inexpensive ponds be constructed on a suitable location. The services of Agent Walker for site selection were offered.[104] The answer received by the Bureau was that the company "would work on it."[105] Hetta hatchery operations ceased after the fish of that year were released.

But the main reason for the closure may have been that the Hetta hatchery did not appear to be a success. There were never enough eggs to fill the facility to capacity. Additionally, there were harvest problems at the mouth of the stream in Hetta Cove.

The Hetta Lake system, considered one of the largest sockeye salmon producers in Southeast Alaska, had been commercially fished by the Klawak cannery beginning in 1886. Harvest by this company averaged only 44,000 fish yearly up to 1890, and then about 47,000 fish until commercial exploitation began in earnest in 1896. In four different years the commercial harvest of sockeye salmon at Hetta Cove averaged over 200,000 fish (Fig 31).

The Hetta hatchery was established during this period of bountiful runs. Data are missing from 1900 to 1904, but by the latter years runs had diminished to a fourth of what they had been. The Hetta hatchery did not operate for the next four years, 1904-1907, and the harvests averaged 44,000 fish, with fish being taken three of those years solely by the Klawak cannery. During the nine years from 1908 to 1917, when the hatchery was in operation, reported catch records show a large variation from a high of over 66,000 fish in 1914, when the

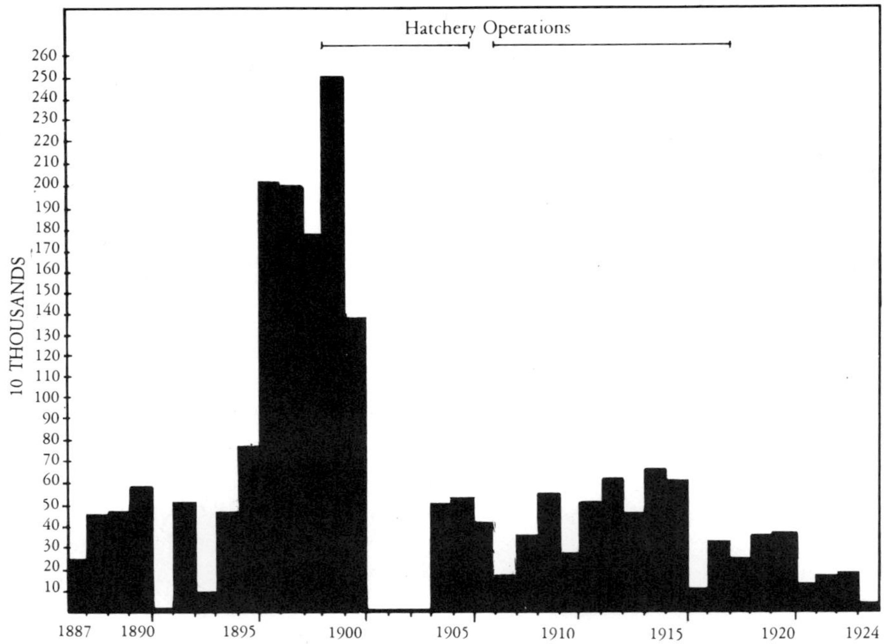

Figure 31. Commercial Harvest of Sockeye Salmon, Hetta Cove, 1887-1924. No data available for 1901-1903. (Source: *Rich and Ball, 1935.*)

hatchery took 7.4 million eggs, to a low of nearly 11,300 fish in 1916, when 3.2 million eggs were taken. The harvest showed continued decreases and in 1924 the entire inlet was closed to fishing.

Fishing gear changed dramatically during the period of the Hetta Inlet fishery. At first haul seines were used. Then purse seines made it no longer necessary to wait until fish schooled near the mouth of the stream. Data show no traps were used in the vicinity until 1912 when three were driven in the Cordova Bay area. That year in Hetta Cove purse seines increased from 8 to 33. For the next ten years, the number of traps fluctuated from one to six; number of seines from 21 to 46; and number of gillnets from zero to fifteen. The majority of the latter two gears fished exclusively at Hetta Cove.

Overfishing of the Hetta Lake system had been recognized as early as 1900 when Moser made his second visit to Hetta. The operators of the hatchery complained frequently that part of their inability to secure adult fish was due to intensive fishing pressure.[106] In 1903 Wisner felt the number of eggs taken was to a great extent interfered with by the fishing in the vicinity of the stream mouth. "The small bay in which the stream empties is nothing more than a pocket of a natural trap," he said.[107] The conformation of the bay led to a high degree of efficiency by seiners. It permitted the laying out of seines in a manner which did not violate the law, and yet effectively deterred considerable numbers of fish from reaching the stream.

Only 2.5 million eggs were secured at the hatchery in 1911, after a dry season, and Fisheries Agent Bower wrote in his annual report:

> "Indications point strongly to the conclusion that overfishing of waters contiguous to the mouth of Hetta Stream had more to do with failure to fill the hatchery than low water. Naturally the latter condition caused fish to loiter longer outside the mouth of the stream than they would had the water been higher."[108]

Fisheries regulations at that time allowed fishing up to 100 yards off the stream mouth at low tide. In 1913 Bower suggested that all fishing be closed from two points, names which no longer are in use and are unidentifiable, but probably designate points at the mouth of the bay. This would provide an area for fish to school in preparation to ascending the stream — an area which in 1913 was fished by up to 16 seiners. Bower wrote:

> "The keen competition of so many different crews means a continuance of fishing throughout the long hours of daylight and its conduct in a manner that will put the utmost number of fish into the boats. Distance restrictions are doubtless worked to the fullest limit of the law. Indeed, it would be very strange if there were not some who were not overly scrupulous in this regard."[109]

It was additionally suggested that, to ensure a good supply of fish to the Hetta hatchery, at least a 500 yard zone off the stream mouth should be set aside where no fishing could take place. A.J. Young, superintendent, suggested that, if commercial fishing were prohibited off Hetta Cove up to July of each season, the June run of sockeyes would be permitted to reach the lake. No action was taken on either proposal. Again in 1914, Bower, in his annual fisheries report, suggested fishing closures if the hatchery was to be able to take sufficient eggs. At that time it was doubted if more than 10,000 sockeyes ever reached the lake.[110]

Finally a hearing was held October 1, 1915 by the Bureau of Fisheries for all interested parties concerned with fishing at Hetta Bay. This resulted in designation of Hetta Bay as a salmon breeding reserve, as stipulated by the 1906 act. Commercial fishing was closed within 500 feet of the stream mouth, as of January 1, 1916. The bay was not posted with markers, which clearly showed the location of 500 yards, until 1920, three years after the hatchery closed. Commercial fishing in Hetta Inlet, from Eek Point to the opposite shore, ceased in 1924.

Subsistence fishing continued at Hetta and exists today. The Alaska Department of Fish and Game maintained a weir at Hetta for four years in the early 1970s and an average of 17,500 fish yearly were counted.

Notes

1. Marshall McDonald. "Report on the Salmon Fisheries in Alaska," *Bulletin of the United States Fish Commission, Vol. XII for 1892* (Washington: Government Printing Office, 1894), p. 3.

2. J.K. Luttrell, "Report of Special Agent Luttrell," Bound in *Seal and Salmon Fisheries, General Resources of Alaska, Vol. II, 1898* (Washington: Government Printing Office, 1893), p. 397.

3. Jefferson Moser, *The Salmon and Salmon Fisheries of Alaska* (Washington: Government Printing Office, 1899), p. 155.

4. Luttrell, p. 396.

5. Jefferson Moser, *Alaska Salmon Investigations in 1900 and 1901,* U.S. Fish Commission (Washington: Government Printing Office, 1902), p. 303.

6. John C. Callbreath, "Diary-1892," (Charles S. Hubbell Papers, University of Washington manuscript collection), October 29.

7. Moser, p. 305.

8. *Alaska Searchlight* [Juneau], October 31, 1896.

9. Moser, *Alaska Salmon Investigations in 1900 and 1901,* p. 308.
Millard C. Marsh and John N. Cobb, *The Fisheries of Alaska in 1909,* Bureau of Fisheries Document No. 730 (Washington: Government Printing Office, 1911), p.14

10. Howard M. Kutchin, *Report of the Salmon Fisheries of Alaska-1903* (Washington: Government Printing Office, 1904).

11. Moser, *Alaska Salmon Investigations in 1900 and 1901,* p. 306.

12. John C. Callbreath, "Diary-1894," (Charles S. Hubbell Papers, University of Washington manuscript collection), April 14.

13. Moser, *Alaska Salmon Investigations in 1900 and 1901,* p. 307.

14. "Hatchery Inspection Reports, 1903-1910," Division of Alaska Fisheries, RG 22, NA.

15. Callbreath, "Diary-1894," January 8.

16. H.C. Fassett, "Memo dated August 5, 1904," File "Hatchery Inspection Reports, 1903-1910," Division of Alaska Fisheries, RG 22, NA.

17. Ibid.

18. Ibid.

19. Callbreath, "Diary-1894," February 6.

20. Ibid, February 24.

21. Joseph Murray, "Report of Joseph Murray, Special Treasury Agent for 1895," bound in *Seal and Salmon Fisheries, General Resources of Alaska, Vol. II, 1898* (Washington: Government Printing Office, 1895), p. 449.

22. *Alaska Sentinel* [Wrangell], December 4, 1902.

23. John C. Callbreath, "Diary-1896," (Charles S. Hubbell Papers, University of Washington manuscript collection), July 26.

24. Ibid, July 27.

25. Howard M. Kutchin, *Report on the Salmon Fisheries in Alaska-1901,* 57th Cong., 1st Sess. Document No. 138 (Washington: Government Printing Office, 1902), p. 54.

26. Moser, *The Salmon and Salmon Fisheries of Alaska,* p. 75.

27. U.S. Congress, Senate, *Report of the Special Agent for the Protection of Alaska Salmon Fisheries,* 56th Cong. 1st Sess. Document No. 153 (Washington: Government Printing Office, 1900), pp. 16-17.

28. Kutchin, *Report on the Salmon Fisheries in Alaska-1901,* p. 54.

29. F.M. Chamberlain, "Report on the Work in Bristol Bay and Elsewhere in Alaska," Reports and Related Records, Division of Alaska Fisheries, RG 22, NA.

30. Ward T. Bower and Harry Clifford Fassett, "Fishery Industries," *Alaska Fisheries and Fur Industries in 1913,* Bureau of Fisheries Document No. 797 (Washington: Government Printing Office, 1914), p. 77.

31. U.S. Congress, Senate, Subcommittee on the Committee on Fisheries *S. 5856 — A Bill to Amend an Act for the Protection and Regulation of the Fisheries of Alaska,* Hearing, 62nd Cong., 2nd Sess. May 20, 1912 (Washington: Government Printing Office, 1912), p. 221.

32. U.S. Congress, Senate, *Report of the Special Agent for the Protection of Alaska Salmon Fisheries,* p. 12.

33. Howard M. Kutchin, *Report on the Salmon Fisheries in Alaska in 1897* (Washington: Government Printing Office, 1898), p. 25.

34. Harry Clifford Fassett, "The Karluk Hatchery," *Alaska Salmon Investigations in 1900 and 1901,* Jefferson Moser (Washington: Government Printing Office, 1902), p. 334.

35. Ibid.

36. Kutchin, *Report on the Salmon Fisheries in Alaska in 1897,* p. 25.

37. J.N. Wisner, ''Alaska Report, 1903,'' File ''Hatchery Inspection Reports, 1903-1910,'' Division of Alaska Fisheries, RG 22, NA.

38. ''Karluk Hatchery, 1903-1954,'' Auke Bay Fisheries Laboratory Records, National Marine Fisheries Service, Federal Records Center, Seattle.

39. Ward T. Bower, ''Fish Culture in Alaska,'' *Alaska Fisheries and Fur Industries in 1911,* Bureau of Fisheries Document No. 766 (Washington: Government Printing Office, 1912), p. 74.

40. H.C. Fassett, ''Memorandum with Regard to the Salmon Hatchery of the Alaska Packers Association on Karluk Lagoon, Kodiak Island,'' File ''Hatchery Data, 1902-1907,'' General Records, Division of Alaska Fisheries, RG 22, NA.

41. Fassett, ''The Karluk Hatchery,'' p. 336.

42. Ibid, p. 341.

43. Howard M. Kutchin, *Report on the Salmon Fisheries in Alaska-1898* (Washington: Government Printing Office, 1899), p. 18.

44. Fassett, ''The Karluk Hatchery,'' p. 343.

45. Ibid, p. 346.

46. John N. Cobb, ''Alaska Diary, 1906,'' (John N. Cobb Collection, University of Washington Archives, Box 4, 11107 C), June 24, 1906.

47. H.C. Fassett, ''Report on the Salmon Hatchery Operated by the Alaska Packers Association on Karluk Lagoon, Kodiak Island, Alaska,'' Reports and Related Records, Division of Alaska Fisheries, RG 22, NA.

48. Fassett, ''The Karluk Hatchery,'' p. 347.

49. Fassett, ''Report on the Salmon Hatchery Operated by the Alaska Packers Association on Karluk Lagoon, . . .'' p. 20.

50. U.S. Congress, Senate, *S. 5856. .* , pp. 234-235.

51. Fassett, ''The Karluk Hatchery,'' p. 348.

52. Kutchin, *Report on the Salmon Fisheries, 1901,* p. 19.

53. Howard M. Kutchin, *Report on the Salmon Fisheries in Alaska-1905* (Washington: Government Printing Office, 1906). p. 22.

54. John P. Taylor, ''Eighteen Months at the Karluk Hatchery,'' *Alaska Magazine,* January, 1964, p. 36.

55. Letter from John N. Cobb to the Commissioner of Fisheries, dated August 18, 1906, File "Karluk Correspondence, 1903-1917," Reports and Related Records, Division of Alaska Fisheries, RG 22, NA.

56. Fassett, "Report on the Salmon Hatchery Operated by the Alaska Packers Association at Karluk Lagoon . . .," p. 3.

57. Ibid, pp. 6-8.

58. Ibid, pp. 22-23.

59. Fassett, "Memorandum with Regard to the Salmon Hatchery of the Alaska Packers Association on Karluk Lagoon . . .," pp. 23-25.

60. Letter from Commissioner of Fisheries to Alaska Packers Association, dated March 17, 1911, File "Alaska Packers Association-Karluk Reports and Correspondence, 1911," Reports and Related Records, Division of Alaska Fisheries, RG 22, NA.

61. Letter from Alaska Packers Association to the Commissioner of Fisheries dated March 20, 1911, File "Alaska Packers Association — Karluk Reports and Correspondence, 1911," Reports and Related Records, Division of Alaska Fisheries, RG 22, NA.

62. Ibid.

63. F.M. Chamberlain, "Report-October 2, 1911," File "Alaska Packers Association-Karluk Reports and Correspondence, 1911," Reports and Related Records, Division of Alaska Fisheries, RG 22, NA.

64. "Memorandum regarding Karluk Hatchery," dated April 16, 1916, File "Alaska Packers Association-Karluk Reports and Correspondence, 1911," Reports and Related Records, Division of Alaska Fisheries, RG 22, NA.

65. Charles H. Gilbert, "Summary of Contributions of the Life-History of the Sockeye Salmon, *Report of Commissioner of Fisheries, 1913,* No. 1 (Victoria, British Columbia, 1914).

66. E.M. Ball, "Inspection of the Karluk Hatchery," Reports and Related Records, Division of Alaska Fisheries, RG 22, NA, p. 1.

67. Ibid, p. 7.

68. Ibid, p. 5.

69. Ibid.

70. "Karluk Hatchery, 1903-1904," Auke Bay Laboratory Records.

71. Ward T. Bower and Henry D. Aller, *Alaska Fisheries and Fur Industries, 1916,* Bureau of Fisheries Document No. 838 (Washington: Government Printing Office, 1917), p. 35.

72. E.M. Ball, "Report on Karluk Hatchery, July 23, 1916," Reports and Related Records, Division of Alaska Fisheries, RG 22, NA.

73. Letter from Alaska Packers Association to Commissioner of Fisheries, dated June 27, 1916, File "Alaska Packers Association-Closure of Karluk-1916," Reports and Related Records, Division of Alaska Fisheries, RG 22, NA.

74. Letter from Alaska Packers Association to Commissioner of Fisheries, dated June 27, 1916, File "Alaska Packers Association-Closure of Karluk, 1916," Reports and Related Records, Division of Alaska Fisheries, RG 22, NA.

75. Letter from Alaska Packers Association to the Commissioner of Fisheries, dated August 14, 1916, File "Alaska Packers Association-Closure of Karluk, 1916," Reports and Related Records, Division of Alaska Fisheries, RG 22, NA.

76. Letter from the Commissioner of Fisheries to Alaska Packers Association, dated March 15, 1917, File "Alaska Packers Asssociation-Closure of Karluk, 1916," Reports and Related Records, Division of Alaska Fisheries, RG 22, NA.

77. Charles H. Gilbert and Willis H. Rich, "Investigations Concerning the Red-Salmon Runs to the Karluk River, Alaska," *Bulletin of the United States Bureau of Fisheries,* Vol. XLIII, 1927, Part II (Washington: Government Printing Office, 1929), p. 7.

78. E.M. Ball, "General Report on the Fisheries of Central Alaska in 1919," Reports of Bureau of Fisheries Agents, Division of Alaska Fisheries, RG 22, NA.

79. Moser, *Alaska Salmon and Salmon Fisheries,* p. 118.

80. Ibid, p. 119.

81. Letter from John C. Callbreath to A.L. Wadleigh dated February 1, 1897 (Charles A. Hubbell Papers, University of Washington manuscript collection).

82. Eggs fertilized August 1 had eye spots appear in 18 days and a few hatched in 44 days, but died. The first surviving fish hatched in 47 days.

83. Moser, *Alaska Salmon and Salmon Fisheries,* p. 115.

84. Jefferson Moser, "Report of the U.S. Fish Commission Steamer *Albatross* from 1st July, 1897 to 31st December, 1897, Part II and Part III," Records concerning the Steamer *Albatross,* General Records, Records of the United States Fish Commission and Bureau of Fisheries, RG 22, NA.

85. John J. Coyle, "Report of John J. Coyle, 1902, Assistant Agent for the Protection of the Salmon Fisheries in Alaska, "Reports and Related Records, Division of Alaska Fisheries, RG 22, NA.

86. Wisner, "Alaska Report, 1903."

87. Letter from H.C. Fassett to Commissioner of Fisheries dated July 26, 1906, File "North Pacific Trading and Packing Company and Hatchery, 1905-1917," Reports and Related Records, Division of Alaska Fisheries, RG 22, NA.

88. C. Wallich, "Memo in re: care and liberation of salmon fry at hatchery at Klawak," File "North Pacific Trading and Packing Company and Hatchery, 1905-1917," Reports and Related Records, Division of Alaska Fisheries, RG 22, NA.

89. Ibid.

90. Ibid.

91. Letter from Wm. K. Hancock to Secretary, Department of Commerce and Labor dated May 4, 1907, File, "Hatchery Data, 1902-1907," General Records, Division of Alaska Fisheries, RG 22, NA.

92. Ward T. Bower and Henry D. Aller, *Alaska Fisheries and Fur Industries in 1915,* Bureau of Fisheries Document No. 834 (Washington: Government Printing Office, 1917), p. 28.

93. Moser, *The Salmon and Salmon Fisheries of Alaska,* p. 115.

94. Wisner, "Alaska Report, 1903."

95. Ibid.

96. Ibid.

97. John J. Coyle, "Report of John J. Coyle, Assistant Agent for the Protection of the Salmon Fisheries of Alaska, Season 1903," Reports and Related Records, Division of Alaska Fisheries, RG 22, NA.

98. *Hetta Hatchery Log, 1910-1918* (Alaska Historical Library: Juneau).

99. Bower, "Fish Culture in Alaska," p. 78.

100. John N. Cobb, "Report of Inspection of Alaska Fish Cultural Stations, Season of 1910," Reports and Related Records, Division of Alaska Fisheries, RG 22, NA.

101. *Hetta Hatchery Log, 1910-1918.*

102. John N. Cobb, "Alaska Diaries, 1910," (John N. Cobb Collection, University of Washington Archives, Box 3, 11107 C).

103. Bower, "Fish Culture in Alaska," p. 78.

104. Letter from Commissioner of Fisheries to Northwestern Fisheries Company dated May 14, 1917, File "Northwestern Fisheries Company Hatchery and Correspondence, 1907-1917," Reports and Related Records, Division of Alaska Fisheries, RG 22, NA.

105. Letter from Northwestern Fisheries Company to Commissioner of Fisheries, dated May 18, 1917, File "Northwestern Fisheries Company Hatchery and Correspondence, 1907-1917," Reports and Related Records, Division of Alaska Fisheries, RG 22, NA.

106. Moser, *Salmon Investigations in 1900 and 1901,* p. 301.

107. Wisner, "Alaska Report, 1903,"

108. Bower, " Fish Culture in Alaska," p. 78.

109. Ward T. Bower and Harry Clifford Fassett, "Fishery Industries," *Alaska Fisheries and Fur Industries in 1913,* p. 75.

110. Ward T. Bower and Henry D. Aller, *Alaska Fisheries and Fur Industries in 1914,* Bureau of Fisheries Document No. 819 (Washington Government Printing Office, 1915), p. 26.

IV. The Mandatory Hatcheries

UNSUCCESSFUL COMPLIANCE

When hatcheries became mandatory for salmon packers in Alaska, after the act of 1900 two major facilities were built: the Fortmann and Quadra hatcheries. Both operated for many years and will be discussed in detail. In addition a few small hatcheries also were started and operated for a year or two in an attempt to comply with the act.

One was on Olive Creek in Olive Inlet, 22 miles south of Wrangell. It was erected by an old time salter, John G. Frey. His hatchery was 20 feet by 16 feet with Williamson troughs and baskets of one million egg capacity. Frey carried on fish culture work in 1902 and 1903. When John Coyle visited on his annual inspection tour in 1903 he wrote that Frey was ". . . a very energetic and successful salmon packer, very much interested in his business, and has established an hatchery on Olive River, where he hatches a large number of cohoes as well as sockeye."[1] Coyle thought Frey might have planted 800,000 sac-fry in 1903. Frey died in 1904, and the man who took over the saltry was not interested in artificial propagation.

Another hatchery which operated briefly was that of the F.C. Barnes Company at Lake Bay, which had first packed salmon in 1901. It constructed and ran a small fish propagation facility in 1902 on an unspecified site in Lake Bay, hatching 850,000 sac-fry before floods washed away the facility. It was too late in the season to reestablish work and plans were made to rebuild the following year. But the cannery was closed in 1903 and, when it reopened in 1904, it was evident few packers were complying with the law and the project was abandoned.

Another hatchery which put fish into the water was built by the Union Bay Packing Company. Although the company built a cannery at Kell Bay on Kuiu Island in 1901, it selected a site at Shipley Bay on Kosciusko Island for its hatchery. Fisheries Agent Coyle found in 1903 that the hatchery, with a bunkhouse and kitchen attached was located on the stream about one-half mile below Shipley Bay.[2] The hatchery was 60 feet by 18 feet and 8 feet high , containing 29 troughs and 174 baskets of the Williamson pattern.

Coyle was told that 1.7 million sac-fry were planted in 1903 by carrying them to the lake.[3] This was the only year the hatchery operated and Coyle was informed that the company did not intend to continue activities. The reason given was that the company ". . . claims that while they plant the fry in the stream, other canneries had equal rights to these fish after they mature and return from the sea and further, the company desires to know, before they went to any further expense, what the government was going to do in view of the Commission [Alaska Salmon Commission] that was sent to Alaska."[4]

Several other canning companies, with an intent to comply with the law, shipped supplies to Alaska and/or actually began construction. All of these attempts were made in Southeast Alaska during 1901, 1902, and 1903. None produced fry.

The most extensive of these unsuccessful attempts was made by the Pacific Packing and Navigation Company which had hatcheries at Quadra and Hetta. It also erected a large hatchery at Freshwater Bay, Chichagof Island, on a small lake above Pavlof Harbor in 1901 (Fig. 32). Wisner in 1903 made a hurried examination and found ". . . a very fine hatchery, considering the building alone. It has been erected on the lake but it has apparently never been used and the water supply which feeds it consists of nothing more than a little seepage which drains through the roots of skunk cabbage and other vegetation."[5]

Insufficient numbers of sockeye went upstream in 1902 to justify putting on a crew to run the facility. The following year Pacific Packing and Navigation Company became involved in financial difficulties and its successor did not list the hatchery among its assets, so it can be assumed the unused hatchery was abandoned.

Three other hatcheries were begun. The Kasaan Bay Mining Company, which built a cannery in 1902 at the village of the same name, began work on a hatchery that year at Karta Lake. The building was nearly completed when it washed away in a heavy rainstorm.[6] Although the intent was to try again the following year, nothing was done. On Annette Island, the Metlakatla Industrial Company purchased materials in 1902 and by 1904 a number of wire baskets were on hand, but nothing further was accomplished.[7] Fidalgo Island Packing Company in Ketchikan built a hatchery on Ketchikan Creek in 1902, but found the situation unsuitable and intended to move the hatchery to a better site the following year. This apparently did not materialize.[8]

Commercial attempts at artificial propagation by private individuals not in the packing business were discouraged by the Treasury Department during this period. It was frequently reiterated that the privilege of the sole use of a stream for hatchery purposes was limited to those engaged in harvesting salmon.[9] On two occasions hatcheries were begun before the individuals learned of this policy. Few details are available on a proposal by H.J. Larrison of Juneau, who cut a trail to an unidentified stream he called Shikt Creek on a lake somewhere near Eagle River. He began a structure near some natural ponds before permission to continue was denied.[10]

Another proposed hatchery was carried much further. A small building, 20 feet by 24 feet, was constructed in 1902 near Juneau on Auke Creek, a short distance upstream from tidewater (Fig.33). It was fitted with 70 baskets, each with a capacity for eggs from 20 salmon. Federal officials did not receive applicaton for the erection of this structure until 1903. At that time Colonel William Winn and Martin A. Needham were requested to cease activities. The

Figure 32. Freshwater Bay hatchery, constructed 1901 but never operated. *Bureau of Fisheries.*

Figure 33. Auke Creek hatchery, constructed 1902 but never operated. *Cobb collection, University of Washington.*

Secretary of the Treasury decided that the application appeared to be in furtherance of a speculative enterprise and would impair the rights of other people.[11]

Despite the denial and instructions to the applicants that all obstructions of Auke Creek would be in violation of the law, when the assistant fishery agent, John N. Cobb, visited the site in 1904, he found a dam with a fishway in the stream about 100 yards below Auke Lake. Cobb asked Winn why private individuals wanted a salmon hatchery. He was told that the owners expected to obtain exclusive rights to fish ascending the stream, or to sell the hatchery to the government, or to sell fry to canneries and saltries which did not operate hatcheries required by law.[12] The dam was removed in 1905 and an unsuccessful attempt to sell the hatching equipment to the Yes Bay Hatchery was made in 1905.

FORTMANN HATCHERY

One of the successful hatcheries to be constructed because of the 1900 act was that of the Alaska Packers Association (APA) on the Naha River system in Southeast Alaska. The literature has identified it by several names, the most popular being the Loring hatchery because it was located near the cannery village of Loring. Then in 1903 the cannery officially named it the Fortmann hatchery for Henry Fortmann, APA president. Thereafter APA and the Bureau of Fisheries called it by the Fortmann name.

The APA cannery at Loring was one of the company's largest operations in Southeast Alaska and it was logical to select that area, on a sockeye system, to meet the hatchery requirements of the new act. The APA management directed Loring Superintendent J.R. Heckman to find a site. He accomplished this in September 1900 with the aid of a U.S. Fish Commission party from the *Albatross,* which included Moser who was surveying the Naha Lake system and enumerating salmon, especially sockeye populations. It was the opinion of this party that the head of the third lake, soon to be named Heckman Lake in honor of the cannery superintendent, would be an excellent site for a hatchery.

Heckman then asked John C. Callbreath, considered the leader in Southeast Alaska fish culture, to make the final selection of the hatchery site. Callbreath also provided a superintendent, his right hand man Fred Patching, who had been in charge of Callbreath's McHenry Inlet hatchery for the past eight years. Patching at that time had more Alaskan experience than any other man except perhaps Richardson at Karluk.

The Naha River meanders in and out of a series of lakes which extend up a narrow rising valley (Map 8). Roosevelt Lagoon is a tidal lagoon connected to the head of Naha Bay by a narrow passageway that can be negotiated with shallow draft boats at slack tide. This short, narrow stretch of water is considered part of the Naha River. At the eastern end of the lagoon is the mouth of the river which connects with Jordan Lake,[13] then Heckman Lake, Patching Lake[14] and Chamberlain Lake.[15] The last two are inaccessible to salmon. It was the concensus of Callbreath, Patching, Heckman, and the party from the *Albatross* that the best site was at the highest elevation sockeye salmon could ascend into the Naha Lake system. The spot selected was on a small arm of land at the mouth of the Naha River where it enters Heckman Lake after cascading for about a mile from Patching Lake. It was near the most important natural spawning beds of the sockeye and near what at that time was considered a good source of water.

The APA began work in May 1901 to open a trail between Naha Bay and Heckman Lake, a distance of about six miles. Four portages of varying lengths up to 1.5 miles were constructed of four foot wide planks. Most of the trail traversed deep ravines and necessitated high trestles. During construction all the planks were packed by hand, some requiring two men. In later years a cart

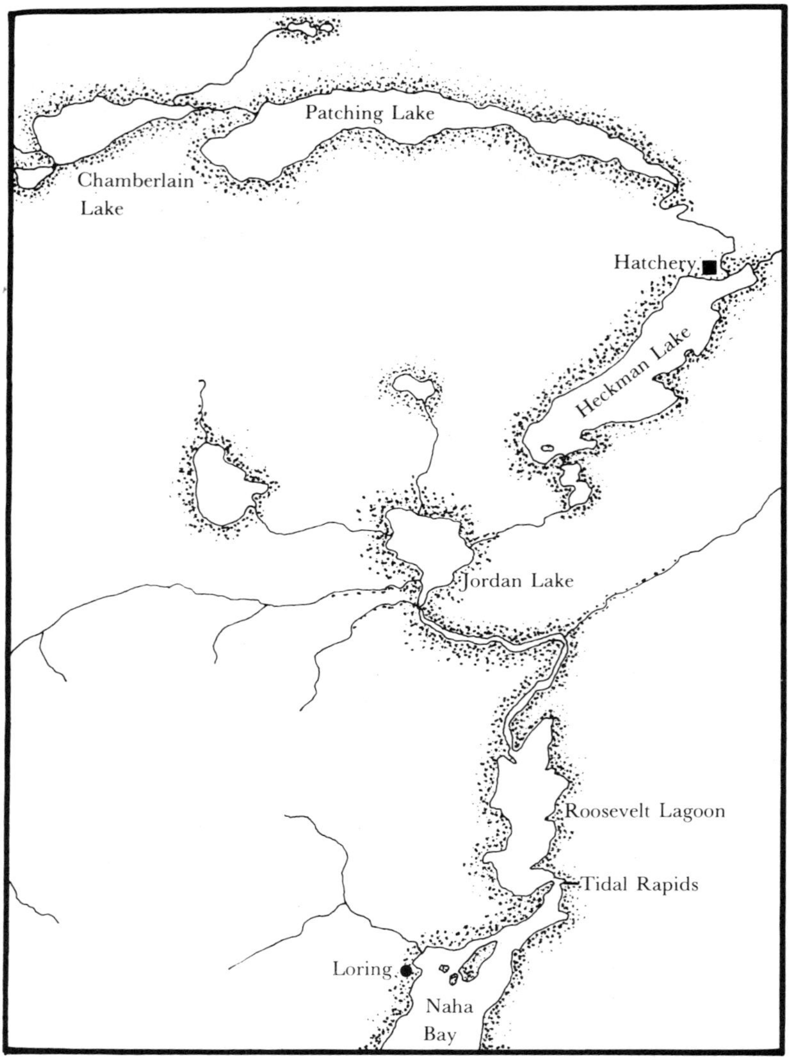

Map 8. Naha River System and Fortmann Hatchery Location

system was developed, at one time pulled by mules. (The latter system was not entirely successful because at least two mules were killed when run over by their carts.) Necessary landings and store houses were completed and boats and lighters placed on the lakes.

Ground was broken for the hatchery as the work progressed on the transportation system. Several years later a sawmill was installed to provide sawn lumber for building materials, but this first hatchery building was mainly of log construction. Men whipsawed lumber and handhewed hundreds of thousands of shakes. The original building, 26 feet by 80 feet, and a cabin, messhouse and quarters were completed October 1, 1901. The first fish had been stripped September 6, and capacity of 11 million eggs was reached that fall.

The hatchery was expanded during the summer of 1902 by an addition bringing the total size of the building to 24 feet by 140 feet, and increasing the capacity to 40 million eggs. Some of the timbers in this new addition measured 100 feet in length and four feet in diameter.[16]

Steam heating and electric light plants were installed as well as necessary outbuildings including dwellings, messhouses, chicken houses, stables and pens for cattle and hogs. The garden, fertilized with fish carcasses, produced vegetables of such quality and quantity that they were featured in newspaper and magazine articles. A telephone line was strung to Ketchikan sometime prior to 1909 to provide communication.

A further expansion began in 1903 when a second building, 50 feet by 270 feet, of squared timbers, was constructed parallel to the first hatchery. This building, completed in time for the 1904 season, brought the capacity to 106,000,000 eggs, reportedly the largest hatchery in the world (Fig. 34). Expansions were made in an effort to comply with the law which required a

Figure 34. Fortmann hatchery complex, 1914. *Bureau of Fisheries.*

release of fry ten times greater than the number of adult salmon taken. The APA believed this enlarged capacity would meet the requirements for their Loring, Wrangell and Pyramid Harbor canneries.

The water supply for the hatchery the first year came from a small stream emptying into the lake above the hatchery, but it was found to be unfit and several springs on the sidehill behind the clearing were utilized by construction of a catch basin and then conveyance of the water by a small flume. These springs, while not furnishing an abundance of water, were sufficient for the single hatchery building. In case of failure, a supply of water could be drawn from a well situated in the hatchery and employing a steam pump with a 360 gallon per minute capacity. This water system worked satisfactorily until 1903 when it was reported that there was a violent embulation of mud in the springs and then a clearing of the water. From that time on the mud flow would reappear at intervals.[17]

This water problem, together with the demand for a larger supply of water with the increased capacity, caused APA to seek other water sources. A spring bordering Patching Lake was carefully examined and measured and found to flow about 300 gallons per minute of pure water at a temperature of 43 °F in the summer and 40 °F in the depth of winter when the air was 1 °F below zero. A basket of eggs was placed in the spring in order to test the hatching qualities.

The spring was situated 134 feet above and 3.5 miles from the hatchery and cost estimates to bring the water to the hatchery were rumored to be $10,000. The decision was made to connect the spring water despite this excessive cost. Work commenced in 1904 on construction of a tramway 5,524 feet long and four feet wide for the transportation of building materials, including 18,450 feet of eight inch and six inch wooden pipe. Two thousand feet of trestle work, some of it 30 feet high around rocky bluffs, was erected to carry the pipe which was laid in boxes packed with moss and earth. A bridge and trestle 300 feet long was constructed to cross the river, and the divide between Patching and Heckman lakes was passed by tunnel and a cut 22 feet deep.

The new system proved adequate in combination with back-up water supply systems. The electrical power system could be used since the water could be diverted from a flume 1,900 feet long, 3 feet wide and 18 inches deep that ordinarily ran an overshot wheel 23 feet in diameter. A steam pump could bring water from the river if all else failed. All of these systems were necessary in 1909 when 87.5 inches of rain fell between July and November, and large slides from the mountain carried away a considerable portion of the pipeline. The flume provided water during repairs until it too washed away. Water was then pumped from the river.

The arrangement of the troughs caused a peculiar feature of the water system at the Fortmann hatchery. Since the troughs were placed end to end, in a series of 13 troughs in the old wing and 15 troughs in the new wing, water passed in at the head and flowed through 13 troughs in one instance and 15 in the other. Ordinarily in hatcheries built at the turn of the century, water was not conducted through more than three or, at the most, four troughs. Additionally a much smaller flow of water was maintained than at other hatcheries and with the length of the troughs, the number of gallons per minute per thousand eggs was further reduced. There was a fall of eight inches between each two troughs for aeration.

There were 104 troughs in the old wing and 150 in the new wing which at seven baskets to the trough, gave a total of 1,778 baskets. Each basket held over 60,000 eggs, thus making the capacity 106,000,000 eggs, although the capacity is often given at 110,000,000 eggs.

The hatchery was located immediately adjacent to the spawning grounds, and fish were captured by means of a haul seine, 35 fathoms long. No mention of a weir on the Naha River near the hatchery was found. It appears that all fishing was done within the river, up to the base of a small dam used to divert water for the power flume. Some seining was done at two small tributaries at the head of Heckman Lake. One tributary, Fence Creek, had a weir.

High water hampered much of the early-day operations. There were numerous years when rain fell during the spawning months so that the river was in constant flood and seining operations were severely hampered (Table 15). An extreme example occurred in 1917 when 78 inches of rain and 13.5 inches of snow fell within the normal spawning period, September through November. Early in November the water stood 33 inches over the hatchery wharf and 12 inches higher than ever recorded. It was impossible to seine adult salmon, and the number of eggs taken that year was the lowest in the hatchery's history.

Eggs from other systems were brought into the Naha River on two occasions to supplement low numbers of eggs taken. Sockeye eggs were taken at Kegan and Miller creeks (the latter called Norwiskay stream) in Moria Sound, Prince of Wales Island when fish could not reach Heckman Lake due to a drought in 1903. When the hatchery at Hugh Smith Lake, Boca de Quadra was closed about six million sockeye eggs were taken at Buschmann Creek in 1907. The latter transfer was not successful.[18] In 1912 some eggs were taken from the stream between Heckman and Jordan Lake where sockeye salmon were reportedly bigger than at the head of the lake.[19] Hatchery personnel, however, relied upon sockeye captured at the hatchery spawning grounds to provide eggs all other years.

Sockeye salmon were incubated at all Alaskan hatcheries, but Fortmann was one of the few hatcheries to propagate other species. Propagation of coho eggs took place from 1903 to 1906. These eggs were from Naha River fish except in 1903 when some spawn came from Miller Lake, Moria Sound. Sac-fry were released in the stream beside the hatchery. (Table 16)

Table 15. Sockeye Production, 1901-1926: Fortmann Hatchery

Brood Year	Spawning Period	Number of Eggs Taken	Number of Fry Released	Comments
1901	Sept. 6-Oct. 5	11,460,000	10,300,000	Hatchery capacity. Fry released March 5, 1902.
1902	Aug. 15-Oct. 21	40,050,000	29,005,000	Hatchery capacity. 27,405 fish used. Fry liberated Dec. 9, 1902-April 28, 1903. Heavy loss due to freezing.
1903[a]	Aug. 20-Nov. 4	20,226,000	12,615,000	Drought, only 13,530 fish captured; supplemented with 3,390,000 eggs from Kegan Creek and 3,306,000 from Miller Creek. First fry planted Jan. 27, 1904.
1904	Aug. 23-Oct. 15	63,120,000	62,160,000	High water prevented seining.
1905	Aug. 11-Nov. 12	68,715,000	67,634,000	High water prevented seining.
1906	...	105,420,000	80,946,000	18,000,000 eggs died when water froze.
1907	...	41,250,000	33,920,000	Poor run. 6,000,000 eggs taken at Buschmann Creek.
1908	...	24,465,000	22,785,000	...
1909	...	53,340,000	50,725,000	High water prevented seining. Fry fed in two ponds and in troughs.
1910	Aug. 26-Nov. 6	34,920,000	30,245,000	Many fish in lagoon, but did not reach spawning beds. 22,000,000 fry fed in ponds and troughs and released into hatchery stream August 1.

Table 15 — Continued

Brood Year	Spawning Period	Number of Eggs Taken	Number of Fry Released	Comments
1911	Aug. 26-Oct. 22	107,520,000	100,335,000	Many fish left after hatchery capacity reached. 40,425 males and 44,266 females used.
1912	...	23,160,000	20,800,000	Poor run. Seined streams between Heckman and Jordan lakes.
1913	Aug. 24-Oct. 16	9,480,000	8,700,000	Poor runs all over Southeast.
1914	Aug. 22-Nov. 30	22,500,000	20,800,000	Poor runs, high water hampered seining.
1915	Aug. 21-Nov. 20	26,520,000	25,055,000	...
1916	Aug. 22-Nov. 20	62,580,000	57,405,000	...
1917	Sept. 3-Oct. 20	6,840,000	6,135,000	Excessive rain (100 inches between July 1-Nov. 30) made it impossible to obtain spawn. 78 inches fell during normal spawning period.
1918	Aug. 28-Nov. 15	19,620,000	15,205,000	...
1919	Sept. 8-Nov. 22	18,420,000	17,070,000	...
1920	Sept. 4-Nov. 11	18,240,000	17,375,000	...
1921	Aug. 26-Nov. 21	13,380,000	12,885,000	...
1922	Aug. 28-Nov. 4	17,760,000	16,985,000	...
1923	Sept. 4-Nov. 24	15,480,000	13,875,000	...
1924	Aug. 22-Nov. 18	11,640,000	11,005,000	...
1925	Sept. 5-Nov. 16	16,920,000	15,990,000	...
1926	Aug. 24-Nov. 18	21,420,000	18,830,348	...

(a) Numbers in published reports include coho.

Table 16. Coho and Pink Salmon Production, 1903-1926: Fortmann Hatchery

Brood Year	Species	Number of Eggs Taken	Number of Fry Released	Comments
1903	coho	1,977,000	1,165,000	Combined lots of eggs from Heckman Lake and Miller Creek (Moria Sound). All released in Heckman Lake or Naha River beside hatchery.
1904	coho	1,890,000	1,021,000	Eggs taken Oct. 17-Nov. 15 at Heckman Lake.
1905	coho	2,424,000	1,873,000	Eggs taken Oct. 3-Nov. 12 at Heckman Lake.
1906	coho	30,000	27,000	...
1913	pink	5,000	...	Eggs from Jordan Lake fish, released fry in Heckman Lake. Return of 119 fish in 1915.
1915	pink	600,000	560,000	Eggs taken Aug. 22-Nov. 22 from returning fish.
1917	pink	2,400,000	1,845,000	Eggs taken Sept 3-Oct. 9.
1918	pink	3,660,000	3,235,000	Eggs taken Sept. 13-Nov. 15.
1919	pink	600,000	...	...
1920	pink	360,000	...	...
1921	pink	900,000	830,000	...
1922	pink	240,000	220,000	...
1923	pink	1,200,000	1,150,000	...
1924	pink	900,000	805,000	...
1925	pink	4,680,000	4,181,000	...
1926	pink	6,640,000	...	Released 756,000 in Nov. 1926 and 2,490,000 in Dec. 1926.

The practice of propagating coho salmon ceased when it became the accepted belief that coho fingerlings actively preyed upon sockeye fry. Dr. Charles H. Gilbert found that coho fry stayed in freshwater for a year and hence to be of value, any coho output of a hatchery had to remain in lakes and streams with sockeye fry. After learning this, Fortmann hatchery personnel ceased coho salmon incubation and went to far as to prevent coho from spawning naturally in Heckman Lake.

Pink salmon eggs were taken in later years. Unlike coho, no damage would be done to the sockeye output since the fry left fresh water immediately. The Naha system supported a large natural run of pink salmon, but the greater portion did not reach Heckman Lake. Moreover, no rebate was received on pink fry released.

All species used at the hatchery were seined and the unripe females returned to the lake. Ripe females were stripped into pans and eggs impregnated with milt, generally on a one female to one male basis, if possible, using either the wet or dry method. It was not until 1917 that the incision method of taking eggs was put into practice. This came about as a result of an inspection by a fisheries agent who found, in July 1916, that "The incision method of taking eggs is not practiced at this hatchery, nor the fish killed before stripping. Upon examination of the pocket between the gills, a number of females had 100 eggs average per fish."[20] The Bureau of Fisheries requested that APA instruct its employees to take eggs by incision to eliminate egg waste. After that time, this method of taking eggs was put into effect.

The fertilized eggs were placed into baskets, picked only once, and left undisturbed, a procedure not in general practice in other hatcheries. Chamberlain observed this practice of the resting period, or as he called it "burying the eggs during the tender period,"[21] on his visit to Fortmann hatchery. The practicce was still in use in 1925 and as a fisheries agent described it ". . . the eggs receive but one picking after being taken, then are left fifty days without attention. Some which were covered with sediment were in good shape when shook up. The water must be ideal to carry eggs without fungus for that length of time."[22] When the tender period was over, the eggs were picked by hand. The salt solution never was adopted.

Records of the incubation period are scattered and incomplete. On his 1903 visit Wisner collected the following data based on water from the nearby spring source: "Eggs taken September 6, 1903 eyed in 35 days at 535 t.u. and began hatching in 110 days at 1289 t.u. The eggs taken August 15, 1902 eyed in 35 days at 544 t.u. and hatched at 1164 t.u."[23]

The amount of time and temperature required to hatch sockeye salmon in 1905 using the new water system from the spring by Patching Lake were reported as "First take of eggs [August 12], 106 days at average temperature of

43 °F. 1197 Thermo. Intermediate take of eggs, 127 days at 40 °F. or 1177 Thermo. Last take of eggs [November 12], 134 days at 39 °F. or 1153 Thermo.''[24]

Survival from green egg to sac-fry was ''a remarkable record if true,'' according to Willis Rich after a visit in 1923.[25] Patching told Rich the loss ran between two and ten percent.

The sac-fry, as in most Alaskan hatcheries during the early years, were planted in the lake or the Naha River beside the hatchery. A dammed pool in a side stream behind the hatchery received some sac-fry. In some years the sac-fry were packed in buckets, suspended upon a yoke, to the other tributaries at the head of Heckman Lake including Fence Creek. An undetermined number of sockeye sac-fry were placed in the connecting stream between Chamberlain and Patching Lakes in the spring of 1903. Chamberlain, the scientist, found in 1904 or 1905 that Patching Lake was well stocked with cutthroat and Dolly Varden and ''. . . seems to contain also the dwarf form of the sockeye.''[26] The lake today still has a resident sockeye population.[27]

Construction of nursery ponds began in 1903. Two rows of walled ponds fed from the stream varied in size from 16 by 30 feet to 20 by 40 feet. The completion date of these ponds is unknown. When Fisheries Agent Cobb inspected the hatchery in 1910 he reported that Patching, the manager, had 22 million

Figure 35. Rearing ponds, Fortmann hatchery, 1914. *Bureau of Fisheries.*

fry in two ponds of one-half acre apiece. A more complete description of the ponds was made in 1913 by Chamberlain during an inspection trip.[28] He found two large ponds in use, undoubtedly the same ones which Cobb had seen, and two more were under construction (Fig. 35). The bottoms of the ponds were hard earth and the sides were of squared timbers. The bottom of the pond had boulders strewn about to form shelter for the fish. Ponds were cleaned each season by washing the boulders and sweeping and flushing the sides and bottoms. Fry were placed in these ponds sometime after swim-up and fed as long as they remained in the pond.

The first recorded instance of feeding fry in these ponds was in 1909. Cobb was told by the manager that fry in 1909 fed eagerly on herring eggs. The second year they would not take the eggs and were fed fresh mashed steelhead pulp, 30 to 40 pounds each day per 10 million fry.[29] The 11 million fry were fed in 1909 until August 1. Other food may have been tried besides herring eggs and steelhead flesh. John Olafson, who worked at the hatchery in 1909, remembered that livers of fish were cooked, mashed and fed to the fry.[30] Little is recorded after this on the numbers of sac-fry placed in these ponds or if the fry continued to be fed.

Efforts to protect sac-fry and fry from predators were instituted. Chamberlain in 1913 said of the predator control program: ". . . in this regard a more persistent effort seems to have been made to rid waters of enemies than at any other station."[31] A regular campaign against what was thought to be the various enemies of salmon and destroyers of eggs started as early as 1903 when a gillnet was used to catch Dolly Varden. In 1907 wire netted traps were baited with salmon eggs and set in the stream near the hatchery. During the season a number of young trout were captured as well as 40,000 sculpins — 2,700 in one night. These traps were still in use in 1913. Another practice was to dynamite pools to free them of predators before release of fry. For years a hunter was employed to regularly kill and frighten away ducks which were believed to devour eggs and fry.

The APA, particularly J.R. Heckman, welcomed and extended hospitality to professors, scientists, and government officials who came to Fortmann hatchery, with the result that many of the early studies and analyses of sockeye salmon life histories and their culture were conducted at this hatchery. A Bureau of Fisheries shore station was set up at Fortmann in 1903 and 1904 by Chamberlain. He was to reconnoiter the neighboring basins for favorable hatchery sites, observe the methods employed in taking fish at canneries, inquire into the sea habitat and factors influencing the return of adult fish, and inquire into the efficiency of the hatcheries then operated. The publication resulting from this research is Chamberlain's *Some Observations of Salmon and Trout in Alaska.*[32]

Chamberlain conducted outmigrant studies at the point where the Naha River enters Naha Bay, counting 2,314 sockeye fingerlings in 1903 and 8,395 in

1904. His publication explains the variables but does not relate these counts to hatchery production. He also studied the food sources available to the sockeye fry in Heckman Lake. He then repeated the study in 1911 to determine if the food sources in the lake might be overtaxed by the heavy numbers of fry. Measurement of some 80 yearlings from the Naha River in 1911 revealed ". . . the astonishing fact that they were in better condition and larger than fish of corresponding age measured in 1903 and 1904."[33]

Another study carried out by Chamberlain during this period was the most extensive marking program of all the Alaskan hatcheries prior to the 1920s. Results of this study have been discussed in Chapter 2. His studies, although designed for other purposes, did prove that some adult salmon returned to the hatchery.

An innovative experiment was attempted at the recommendation of another of Fortmann's many visitors. Dr. Henry B. Ward, a parasitologist, then of the University of Nebraska and later of the University of Illinois, requested and received money from the Bureau of Fisheries to visit Alaska to research salmon parasites. He spent much time during the summers of 1905 and 1909 at the Fortmann hatchery.[34]

Ward interested Heckman and Patching in propagating plankton as food for salmon fry. This was patterned after a practice in Central Europe where various methods of manuring ponds were in use for many years by the more progressive carp culturists. By 1908 an attempt had begun at Formann hatchery to multiply small organisms in manure-filled trenches connected with the lake. A number of cattle were kept at the hatchery to provide the necessities for such an experiment. A pond was under construction in 1911 to replace the trenches. When Chamberlain visited in 1911 he reported:

> "I understand that the pond was to be fertilized with cow manure and this was expected to stimulate the rapid production and multiplication of certain crustacea. The superiority of this few 1,000 square feet of water over the several square miles naturally available is presumably to be found in the protection of the fingerlings from predatory animals. It can hardly be in the greater quantity of food produced. The experiment will be interesting and should be carefully followed."[35]

This is the last mention of the experiment in any of the surviving literature on the Fortmann hatchery. Principles of pond fertilization with manure were discussed years later in 1928 by George C. Embody in his report of the Bureau's Division of Aquatic Biology and Physics to the 58th American Fisheries Society meeting, but no mention is made of this early Alaskan attempt.[36]

A concern for the technology necessary to make the Fortmann hatchery a success is shown throughout much of the literature of the Bureau of Fisheries, but what were the actual results? An examination of the number of sockeye eggs taken shows that there is such an extreme fluctuation that it is difficult to make any analysis on this basis (Table 15). At one extreme it was possible to take more than 100 million eggs two different years. The annual report of the APA for 1911 gives a description of the large numbers of sockeye salmon available that year, ten years after hatchery operations began:

> "The full capacity of the hatchery was easily filled, and so numerous were the spawning salmon after the hatchery was filled that all the entering streams and lake spawning beds had all the sockeyes that could be massed over them. These spawning salmon even crowded into the flume used to convey power and entered the ditch carrying the overflow from the nursery ponds. As an example of abundance, one small stream, Fence Creek, where we usually place a fence to hold spawning salmon and where we generally take over 3,000,00 eggs, this year yielded over 43,000,000 eggs, and after the fence was removed such a large number of salmon entered that every foot of available spawning ground was covered with spawning fish."[37]

Other years there simply were not enough returning fish on the spawning grounds or high water prevented capture of adult salmon. During the last years of the hatchery operations, it was no longer the policy to explain external influences on the number of eggs taken, so it is impossible to ascertain whether fish were abundant.

Barton Evermann, chief of the Alaska Fisheries Service of the Bureau of Fisheries, made one of the earliest analyses of the hatchery's efficiency by comparing the number of eggs taken from 1903 to 1911. He ignored the influences mentioned above:

> "Until 1911 all the fish that could be secured in the upper principle lake were taken for hatchery use, so an examination of the hatchery records furnish an approximate idea of the run. The year 1906 marked the first big impulse, but contradictory to the theory of building up an isolated system by attention to it exclusively, the years following 1906, (except 1911), show a smaller take than those proceeding. For example for the first years 1903, 1904, 1905, the average take of eggs was 52 million, whereas in 1907, 1908, 1909, and 1910 the average was only 38 million. Thus, it would seem that artificial as well as natural production has failed to elevate the output of an isolated stream above the normal level for the region."[38]

Another analysis based upon numbers of eggs taken was published by Fred M. Chamberlain and Ward T. Bower using data to 1912. Like Evermann they assumed that the entire number of sockeye salmon entering Heckman Lake were artifically spawned. They stated:

> ". . . the number of fish may be arrived at approximately by omitting the last three places in the figure for eggs [taken each year]. For example in 1912, 23,160 fish may be credited to the stream [see Table 15]. From 1887 to 1891 this stream yielded to the canneries an average annual catch of 78,000 fish; in the next five years it dropped to half of that, or 39,000; in the next four years to less than half of this, or 15,000. Under hatchery operations on a closed stream, the run the first four years averaged annually about 33,000; in the next four years it reached 60,000, but in the last four years [1909-1912] it has fallen again to an average of 54,000. In the initial period of four years [1901-1904], of course, only the natural run was produced; that is the product of the hatchery had no influence on the number of adult fish in the stream. The first two seasons not all the fish were spawned and the average, 33,000 is too small. It is probable that 40,000 would be more nearly correct."[39]

Using combined numbers from the Naha and Yes River, Chamberlain and Bower came up with the calculation that an approximate figure of 500 hatchery fry were produced to bring back one adult sockeye salmon.[40]

Only one comment was located regarding the pink salmon returns from hatchery-produced fry. The 1915 annual report of the APA stated that 5,000 pink eggs were taken from fish collected in Jordan Lake in 1913. Fry resulting from the eggs taken were liberated from the hatchery into Heckman Lake where practically no pink salmon spawned naturally. In the fall of 1915, 85 females and 24 males, or 119 pink salmon were reportedly caught on the hatchery grounds.[41] Since the majority of the records of the Naha River system are for sockeye salmon, it is difficult to tell how successful the pink propagation was during the 12 years it was practiced. An examination of table 16 gives an impression that in some years there was sufficient spawn from returning fish for propagation, and that some years supplemental eggs from other sources were used. Alaska Department of Fish and Game records do not include the strength of the pink salmon run in Heckman Lake today.

A fairly good record of the commercial harvest of the Naha River sockeye run is available. This run has been fished since early times. A saltry was built at the outlet of the lagoon in 1883, and in 1888 a cannery was built at the present site of the village of Loring. The Naha River, one of the better sockeye salmon streams in Southeast Alaska, suffered from use of barricades in those early years. The barricade was placed at the head of Naha Bay. A trap in 1893 had been driven almost entirely across the bay, but as few fish were taken, more piles were driven, and it is said a close-mesh net was stretched across the stream so effectively that not a salmon could pass.[42] Fishing at first was carried on with drag seines, one crew fishing in Naha Bay below the tidal rapids and the other in Roosevelt Lagoon.

Moser, during his examination of sockeye streams in 1897, made the following annual harvest estimates: "The best authorities consider the stream good at present under fair conditions for 50,000 redfish [sockeye], 5,000 cohoes, and 400,000 humpbacks [pink], and if properly cared for, it should yield 75,000 redfish per year."[43] Chamberlain also collected catch data and this is shown in figure 36.

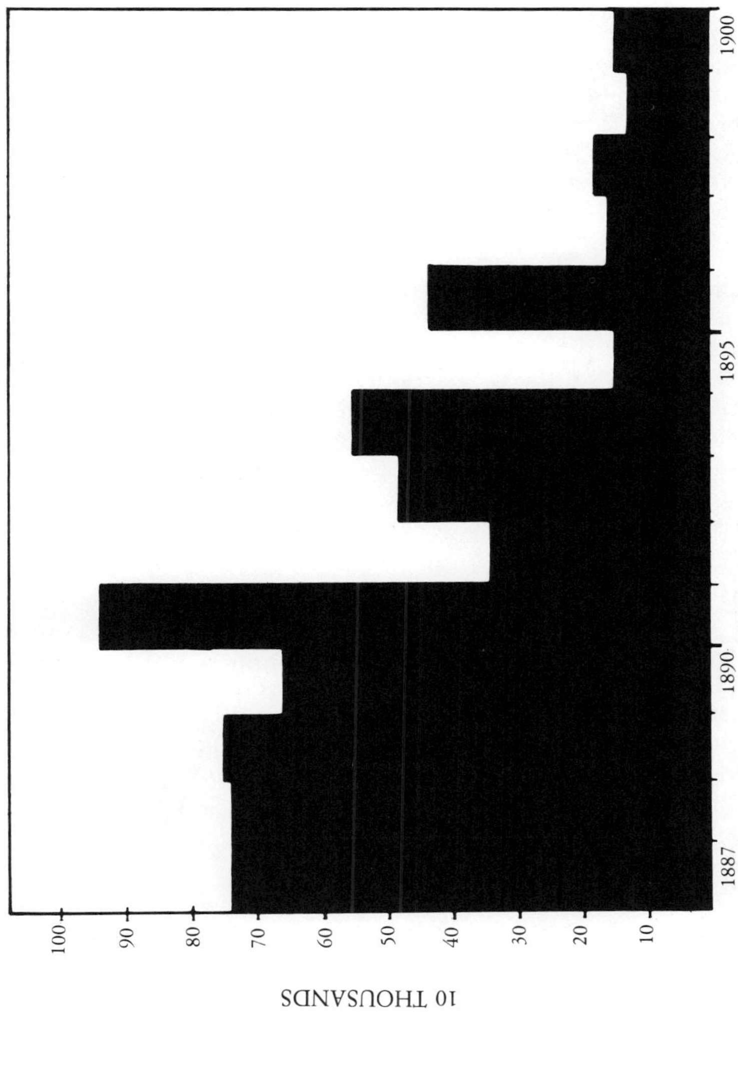

Figure 36. Catches of Sockeye Salmon, Naha River, 1887-1900. (Source: *Chamberlain*, 1907.)

These data show the downward trend prior to construction of the hatchery in 1901. Chamberlain explained that the large catch in 1891 was due to close barricading as well as to an abundance of fish. He noted that, while in a general downward trend, declines were most marked in four year periods after that date. The run was poor in 1903 when only 20 million eggs were taken, and again in 1907 the run did not meet expectations. But 1911 was a phenomenal year when 84,691 sockeye salmon were used for spawning and more were left in the stream. Chamberlain made one further observation on his catch data: "Conclusions based on the commercial catches, however, require not only accuracy in the statistics but details as to the efforts made to take all fish in the given waters. Yet these figures undoubtedly indicate the possibility of fishing out a given stream."[44]

As the runs declined, the necessity for regulations became obvious. In 1889 barricades first became illegal. Then, in 1901, fishing was restricted to 100 yards from any stream mouth. This did not seem a sufficient measure to protect the Naha River system so it was closed to fishing from January 1 to July 15 and from October 15 to December 31 of the year 1902. The APA officials felt that by waiting until July 15, the majority of the sockeye would be in the system. This order, however, was not repeated and again in 1903 the only regulation in effect was the 100 yard limit.

It became obvious to one fisheries agent, John Coyle, that the regulations were not being observed. On his way to the hatchery in 1903 for an inspection, he rowed through Roosevelt Lagoon and found fishing crews of the Loring and Yes Bay canneries. He proceeded to evict them from the lagoon, reminding them that by decision of the U.S. Treasury Department (then in charge of Alaska's fisheries) the Naha River was closed to fishing. As a result Coyle reported "Mr. Fortmann, the president of the Alaska Packers Association afterwards took issue with me, claiming that by the peculiar wording of the law under which the regulation was made, they were permitted to fish because they were operating a hatchery of sufficient magnitude to keep said stream stocked."[45]

It was many more years before further attempts were made to protect the Naha River. A hearing was held before the Bureau of Fisheries in 1912 to consider closure of the Naha River stream, its catch basins frequented by salmon, and an area or zone within 500 yards of the mouth of the stream. During the hearing unreported illegal fishing episodes were brought to light by Fred Swanson, an agent of the Alaska Fishermen's Union. He testified that fishermen sent by canneries went into Roosevelt Lagoon and brought fish out "going so far as to haul their net, throw out the pink and take the sockeye."[46]

By this time Moser was employed by APA and, as their spokesman, his testimony reflects a change in attitude on the part of the company:

> "It is our desire to conserve the stream as much as possible, not only for the purpose of continuing the hatchery as far as the law permits, but also to prevent a large gathering of fishermen, including

ourselves, within the narrow bay where probably a good deal of destruction might be committed . . . . The configuration of the bay is a tunnel shape and the stream is of such a nature that if seine fishing is permitted within these narrow limits, and the hatchery returns come in as we hope they will, eventually in large numbers, there will be very little chance of the fish getting up the river.''[47]

The Bureau of Fisheries decided in 1913 to close the bay to a line connecting Loring Point with House Point. The entire bay was later closed by order of June 21, 1912.

But the temptation to fish was always there, and it is not known how many years the creek was robbed. A recorded instance occurred in 1921 when a seine boat and crew were found fishing ''within 500 yards of the mouth of Naha Stream near Loring'' but were actually inside Roosevelt Lagoon.[48] They were brought to trial at Ketchikan before the United States Commissioner. Upon pleading guilty the crew was fined $10 each, without costs, while the master of the ship was fined $25 and costs of $9.20.[49]

Another infringement occurred in 1922 and so enraged a private citizen who witnessed the fishing violation that he wrote to Alaska's governor. The citizen wrote:

''What happened while I was staying at the hatchery was this: A seine boat came into the lower lake, or what is known as Salt Lake, on the Naha River, and fished with their seines at the head of Salt Lake where the salmon were laying in thousands, resting before going upstream to the spawning grounds. Of course no one knows how many salmon they got for as soon as they completed their dastardly work they cleared out of the lake and undoubtedly hurried on to some cannery where they sold their fish. There was no one on the lake to stop this piracy and as the same thing happened in 1921 I would be interested to know why the fish commissioners of Alaska do not see to it that a warden is put there while the salmon are running to prevent such acts.''[50]

Hatchery personnel could not prevent such fishing nor even report it because of the distance from the fishing grounds. There were few Bureau of Fisheries agents in Alaska and the Territorial Fish Commission had no authority to regulate fisheries. With piracy such as this and penalties which did not prevent risk taking, it is safe to speculate that many fish were taken illegally from the Naha River system.

The anticipated spectacular returns to the APA hatchery did not provide Loring cannery with a bigger pack. This undoubtedly influenced the decision to close the hatchery, but another factor actually precipitated the closure. This was the insistence by the Bureau of Fisheries that all sac-fry be held before release. This pressure was prompted by an inspection visit by a fisheries agent who wrote the Commissioner of Fisheries questioning the practice of planting fry in the sac stage in small lakes and ponds and taking credit for fry planted and charged against tax in the form of rebates.[51]

The Commissioner of Fisheries wrote William Timson, APA president, in February 1927 requesting that the company rear sac-fry to the full free swimming stage. "I know," he wrote, "that the law permits a tax rebate for release of red salmon fry [sac-fry] but I wish to recommend that you consider the rearing of these fry along the lines herein outlined before planting."[52] Timson wrote back six days later:

> "As you probably are aware, we have kept fry until they were about 4 or 5 inches long before liberating. In fact, we have tried all sorts of schemes to insure a larger return from the fry liberated but unsuccessfully.
>
> "We have decided it is a waste of time to continue operations at the hatchery and have given instructions to Mr. Patching not to take eggs in 1927. If the Bureau of Fisheries cares to take over the operations of the plant with an idea of experimenting, we will gladly turn it over to them."[53]

The offer was declined because insufficient funds were available to operate it. When it was suggested that the Territorial Fish Commission take over the Fortmann hatchery, the offer was also refused for the same reason. Timson's final correspondence about the Fortmann hatchery said:

> "The writer's personal opinion is that we have been an 'angel' for fish hatchery experiments long enough . . . . We have spent a very large sum of money in building and operating this hatchery during the past twenty-five years. However, with considerable regret, we have concluded that fish cultural operations are of little if any benefit, and therefore we reluctantly consign the Loring Hatchery to oblivion."[54]

QUADRA HATCHERY

Another hatchery built in response to the 1900 act was located at Hugh Smith Lake and operated for a longer period of time than any of the other Alaskan hatcheries (Map 9). It was also the last of the private, rebate-subsidized hatcheries to close. The Quadra hatchery was started in 1900 by Quadra Packing Company. The stockholders in this company were also involved in the Straits Packing Company in Petersburg and it was intended that the Quadra hatchery would supply the required fry for both canneries. The hatchery was

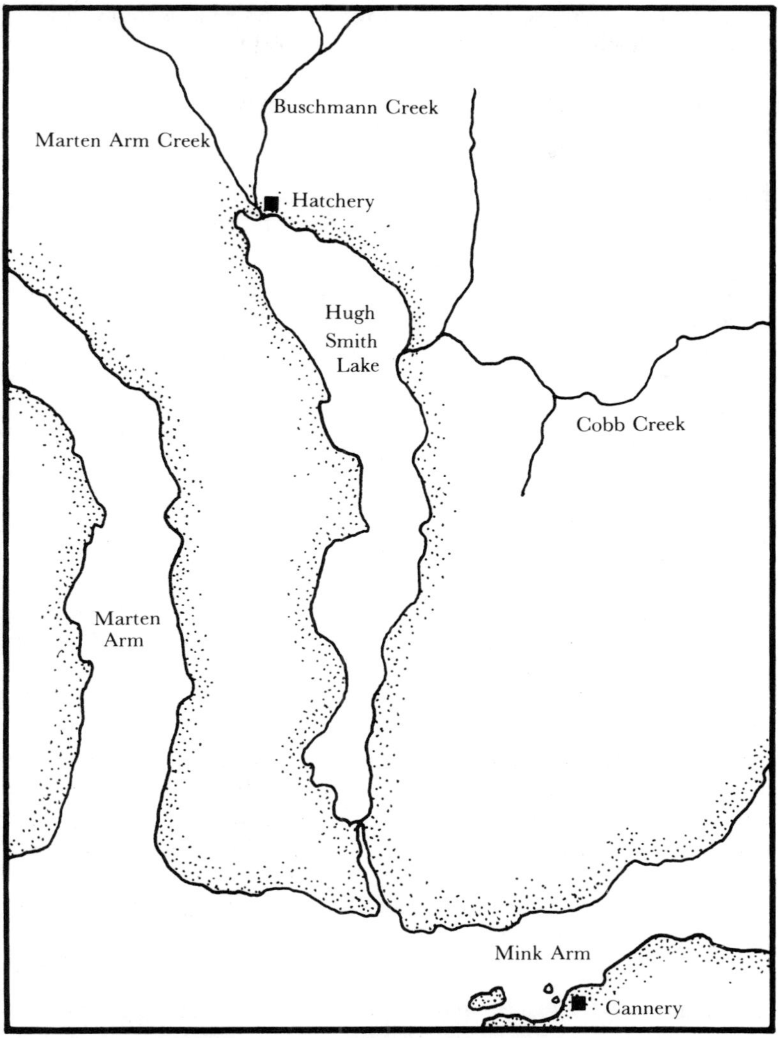

Map 9. Hugh Smith Lake, Boca de Quadra and Quadra Hatchery

operated with little success and was part of the assets sold to Pacific Navigation and Packing Company in 1901. This was just one of the many transfers of ownership of the Quadra hatchery. When Pacific Navigation and Packing Company declared bankruptcy, the hatchery closed (1904-1907). Then Northwestern Fisheries Company purchased some of the properties at the resulting sale, including the Quadra hatchery. The hatchery was reopened and operated until the property was leased to Pacific American Fisheries in 1933. That company ran the hatchery for three years before closing it permanently.

The original hatchery, constructed during 1900 on what is today called Cobb Creek, was located about 150 to 200 feet from the mouth of the creek.[55] The water flowed through a flume of two inch material for a distance of 100 feet to an 18 foot by 70 foot hatchery building. The only report on the activities of this hatchery was made by Wisner in 1903.[56] Operations, he learned, were under the supervision of Emil Glasso, who had worked in Puget Sound hatcheries. He and his father had been hired for the company's hatchery at Freshwater Bay and Emil had been transferred to Quadra to handle the new operation.

After the first unsatisfactory year when only 600,000 eggs were secured, a new superintendent O. Kettlesby assumed charge. He immediately moved the hatchery to a site he selected on a tiny tributary, one-eighth mile long, near the head of Hugh Smith Lake, close to today's Buschmann Creek (then known as Quadra River). [57] This tiny creek, soon known as Hatchery Creek, rose out of a spring, which proved to have the warmest water in the area. Water temperatures, according to Wisner, averaged 40 °F in the winter. The Quadra hatchery logbooks confirm this, although some winters the water averaged 42 °F.[58]

A small reservoir, about 100 feet long and 25 to 30 feet wide, held water from the spring and what was needed in the hatchery ran through a wooden flume. Even during periods of low water the spring provided 50 gallons per minute. The water proved to be excellent and no problems with fungus or disease were recorded.

A building 70 feet by 17 feet contained the hatchery, a mess and sleeping quarters. Standard Williamson troughs with 162 baskets were used with 30 troughs in five rows. Twenty-four were 14 feet long and six were 6 feet long, all 14.5 inches wide, inside measurements, and 7.5 inches deep. The lower trough was only 18 inches from the floor and the highest 4.5 feet from the floor, which sloped in the direction the water flowed. Water was supplied by a head trough 17.5 feet long and 12.5 inches wide, 8 inches deep.

This hatchery was replaced in 1912 to increase the capacity from 10 to 17 million eggs. The new structure, 87 feet long and 24 feet wide, was devoted entirely to hatchery work, and a separate dwelling built for employee quarters (Fig. 37). The equipment for the hatchery consisted of the 30 troughs of two lengths, supplemented by 20 troughs.

Spawning sockeye salmon frequented Buschmann Creek as far as the falls, about three miles upstream. The fish were captured for hatchery use with 60 to 70 foot haul seines at the mouth of the creek. To facilitate this, a rack was constructed 47 feet wide, 6 feet deep with 160 bars of seven-eights inch iron, each eight feet long. High water occasionally hampered effectiveness, and some sockeye salmon escaped to spawn naturally. The rack was pulled at the end of the spawning period.

Figure 37. Quadra hatchery building, 1914. *Bureau of Fisheries.*

A run of sockeye developed in Hatchery Creek, and a trap was built into the creek to capture these salmon. Diversion of water from Buschmann Creek into Hatchery Creek provided adequate water flow to the trap. Superintendent Mark Meyer remembers that by 1934-1935 the adult returns to Buschmann and Hatchery creeks were about equal.[59]

After the fish were seined or taken from the trap, they were tested for ripeness. Some adult salmon, in early years, were held in the fry ponds for ripening, but it is unknown how long they were held. Meyer stated that in 1934 only fish which would ripen in a day or two were held.[60] The green ones were returned to the lake.

Eggs were taken by the pressure method until the season of 1912. That year the first eggs were taken by pressure, and then it was decided to try the new abdominal incision method. It was found, when several females which had been stripped by pressure were examined, that from 100 to 400 good eggs remained inside the fish. That year the hatchery personnel converted to the dry method of fertilizing eggs (stripping into a pan merely moistened instead of into a pan half filled with water). With these improved methods a nine percent loss from green egg to sac-fry was experienced. Disposal of fish carcasses was into the lake, except for a few used as fertilizer in the gardens. Table 17 shows the number of eggs taken yearly at the hatchery.

Table 17. Sockeye Spawn Taken, 1901-1935: Quadra Hatchery

Brood Year	Number of Eggs Taken	Spawning Period	Comments
1901	600,000	...	Hatchery located on Cobb Creek.
1902	4,088,000	Aug. 23-Dec. 17	...
1903	600,000	...	No fish. Blamed on fishing pressure at mouth of Sockeye Creek.
1904	...	...	Hatchery not operated.
1905	...	...	Hatchery not operated.
1906	...	...	Hatchery not operated.
1907	...	...	Hatchery not operated but over 6,000,000 eggs taken for Fort-mann.
1908	3,325,000	Aug. 18-Nov. 12	...
1909	10,863,000	Aug. 11-Sept. 13	Hatchery capacity.
1910	11,200,000	Aug. 10-Sept. 16	Hatchery capacity.Remaining fish spawned naturally.
1911	11,000,000	Aug. 24-Sept. 28	Hatchery capacity.
1912	10,000,000	Aug. 16-Sept. 30	Hatchery capacity.
1913	18,400,000	Aug. 11-Oct. 11	Hatchery capacity after remodeling.
1914	21,300,000	Aug. 5-Oct. 10	Hatchery capacity.Extra troughs added for season.
1915	17,408,000	Aug. 9-Nov. 13	Hatchery capacity.
1916	16,125,000	Aug. 9-Nov. 15	...
1917	13,600,000	Aug. 15-Nov. 3	...
1918	20,400,000	Aug. 8-Sept. 24	Hatchery capacity.
1919	11,710,000	Aug. 14-Nov. 14	Insufficient numbers of fish. Very cold, seine freezing on last spawn take.

Table 17 — Continued

Brood Year	Number of Eggs Taken	Spawning Period	Comments
1920	19,450,000	Aug. 5-Oct. 28	Hatchery capacity. Fish scarce when seining ceased.
1921	9,985,000	Aug. 11-Nov. 14	Low water in Aug. High water several times remainder of season with water flowing over weir. Small run of fish.
1922	6,195,000	Aug. 11-Nov. 19	Floods during much of spawn take period.
1923	17,885,000	Aug. 12-Nov. 7	Hatchery capacity.
1924	20,050,000	Aug. 7-Sept. 30	Hatchery capacity.
1925	20,240,000	Aug. 5-Sept. 9	Hatchery capacity. Estimated half of run used to reach capacity.
1926	20,000,000	Aug. 5-Sept. 3	Hatchery capacity.
1927	20,240,000	Aug. 12-Oct. 13	Hatchery capacity.
1928	20,310,000	Aug. 6-Sept. 13	Hatchery capacity. Estimated as many fish spawned naturally as spawned for hatchery use.
1929	11,760,000	Aug. 12-Nov. 15	High water prevented seining adults. Runs may have provided insufficient fish.
1930	21,190,000	Aug. 2-Sept. 17	Hatchery capacity plus extra troughs.
1931	20,280,000	Aug. 6-Sept. 7	Hatchery capacity.
1932	25,895,000	Aug. 4-Sept. 5	Hatchery capacity plus troughs set up outside.
1933	20,650,000	Aug. 4-Sept. 6	Hatchery capacity.
1934	10,221,000	Aug. 8-Nov. 10	High water prevented adult capture.
1935	34,383,000	Aug. 8-Sept. 24	Hatchery capacity plus extra troughs.

Table 18. Incubation Data, 1913: Quadra Hatchery

Date Eggs Taken	Days to Eye	Days to First Hatch	Days to Last Hatch	Hatching Range
AUGUST				
11	39	104	115	11
14	39	104	114	10
15	39	104	114	10
16	38	105	116	11
18	39	103	116	13
22	39	104	116	12
23	39	104	119	15
24	40	104	118	14
25	40	105	117	12
26	40	105	116	11
27	40	105	117	12
29	41	106	116	10
30	41	106	117	11
SEPTEMBER				
2	42	107	119	12
4	41	108	120	12
5	42	108	120	12
6	42	109	121	12
8	42	110	123	13
10	42	106	121	15
12	42	108	121	13
13	42	109	121	12
15	43	110	122	12
16	43	112	122	10
17	43	112	123	11
18	43	111	124	13
19	43	110	124	14
20	44	111	124	13
22	44	111	124	13
23	44	109	124	15
24	44	112	125	13
27	44	112	125	13
29	44	113	127	14
30	44	114	128	14
OCTOBER				
1	45	114	129	15
3	45	115	131	16
5	45	116	132	16
7	46	116	134	18
8	46	117	134	17
10	46	117	135	18
11	47	118	136	18

Incubation procedures were those commonly in use at the time. The salt method of picking dead eggs was adopted sometime during the 1920s. Incubation data are available in the Quadra hatchery logbooks and an example of incubation periods for 1913 is shown in table 18.[61]

Once the eggs began to hatch, the sac-fry were placed in Buschmann Creek and two of its tributaries, Trap Creek and Marten Arm Stream. The latter must not be confused with streams flowing into Marten Arm of Boca de Quadra. It refers to a stream which is today called the left fork of Buschmann Creek. Sac-fry were never released into the lake.

As early as 1908 sac-fry ponds were in use at the Quadra hatchery. These were built within Hatchery and Trap creeks and can be seen in figure 38. As many as nine of these ponds eventually were in use. They were ponds excavated with pick and shovel, 15 feet wide by 50 to 100 feet long, depending upon the contour of the creek. They had natural gravel bottoms.

On an inspection trip in 1910, Fisheries Agent Cobb describes a pond in Hatchery Creek as being 50 feet by 25 feet and covered with a framework of logs with boughs placed on top to keep out predators (Fig. 39). In later years these covers were replaced with webbing. The ponds were considered a safe refuge closely representing natural conditions. The sac-fry were scattered around the fringe of the ponds so they could hide under the limbs and brush. The lower ponds were planted first and as fry swam up, they would work their way into the lake. The water cooled enough by Thanksgiving that the sac-fry stayed in the ponds until the spring when they worked their way from pond to pond to reach the lake about a thousand feet away. "We could see progressive rates of growth in accordance with the temperature," Superintendent Meyer stated. "We saw the change in length, and in color and could see them maturing and finally migrating to the lake."[62] Table 19 shows the number of sac-fry released into these ponds.

Fry were never fed at the Quadra hatchery. The Bureau of Fisheries agents and the Commissioner of Fisheries began by the late 1920s to urge the company officials to hold sac-fry to the full swimming stage before release into ponds. The Commissioner wrote in 1927 to the company officials:

> "It is my belief that better results would follow from fish culture operations now conducted if the salmon fry were reared to full swimming stage and as much longer as possible before planting. I am sure you will feel as I do that this will result in a larger return of salmon. I appreciate that the question of available pond space and added cost of feeding the young must be taken into account."[63]

The company ordered no change in their operational plans. Then, in November 1929, a fisheries agent wrote the Commissioner of Fisheries after examining the Quadra operations that he felt the pond facilities were very limited

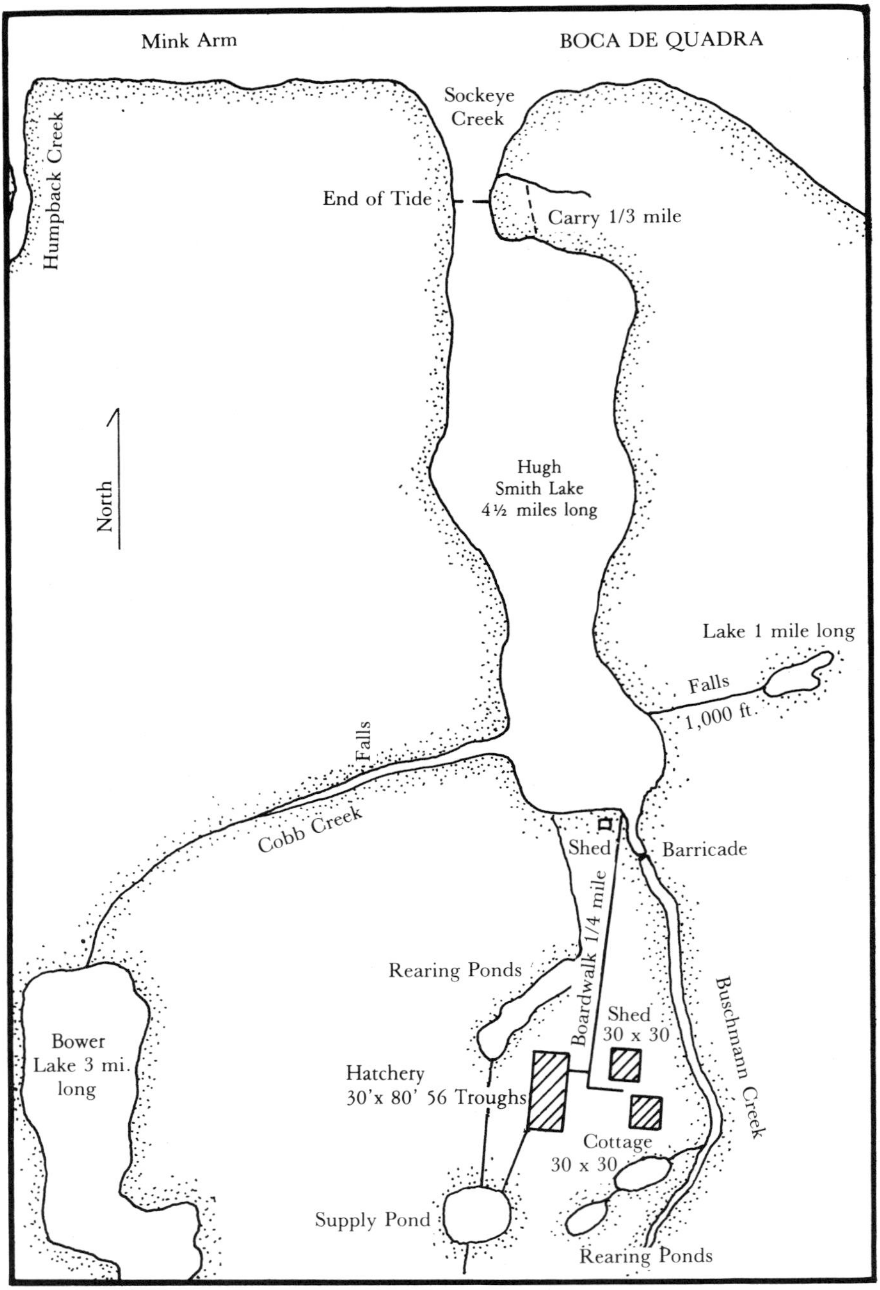

Figure 38. Layout of Quadra hatchery complex, 1914-1935. *Bureau of Fisheries.*

and the sac-fry could easily bunch up and smother if large numbers were released into the ponds at once. Additionally, sac-fry were easy prey for predatory birds and fishes. He suggested installation of a stacked tray system such as the one in use at Yes Bay hatchery. This would allow care of the young fish until the yolk-sac was absorbed. He also recommended a program to rid the waters of predatory trout.

The company apparently ignored the suggestions, although Meyer remembers that someone dabbled with the tray system at some time. The sac-fry continued to be planted in the ponds. There was no organized effort to eliminate predators although wives of some of the employees caught Dolly Varden for the bounty.

The last sac-fry to be planted were those from the eggs taken in 1935. Meyer explained that the hatchery was closed by Pacific American Fisheries after the privately controlled company became a publically owned corporation. Any part of the company which did not make a profit was eliminated. The hatchery was abandoned.

Over the years the various canning companies which operated the Quadra hatchery required no evaluation of its success. There were no outmigrant studies nor outmigrant captures. A few marked fish were reported in 1910 by Cobb. He listed seven such sockeye salmon, six of which were females.[64] Two

Figure 39. Quadra hatchery rearing ponds with covers to exclude predators, 1914. *Bureau of Fisheries.*

Table 19. Sockeye Sac-fry Released, 1901-1935: Quadra Hatchery

Brood Year	Total Release	Date of Release	Comments
1901	...	...	No records survived.
1902	...	...	Last sac-fry released two weeks after hatching.
1903	0	0	Eggs planted in stream.
1904	0	0	Hatchery not operated.
1905	0	0	Hatchery not operated.
1906	0	0	Hatchery not operated.
1907	0	0	Hatchery not operated.
1908	3,025,000	Dec.-Feb.	Released in ponds, Buschmann and Hatchery creeks.
1909	9,850,000	Dec.-Feb.	Released in Buschmann, Marten Arm, Trap and Hatchery creeks.
1910	10,350,000	Dec.-Feb.	Released in Marten Arm, Hatchery and Trap creeks. 4,030,000 in ponds.
1911	10,166,000	Dec.-Mar.	Released in Hatchery and Marten Arm creeks. 7,406,000 in ponds. 500,000 held 20-30 days before release into Trap Creek.
1912	8,127,000	Dec. 7-Apr. 17	Released in Hatchery and Marten Arm creeks. 7,000,000 in ponds.
1913	17,654,000	Nov. 25-Feb. 21	Released in ponds and in three ponds in Trap Creek; some held in troughs.
1914	20,300,000	Nov. 11-Feb. 20	Released in nine ponds.
1915	7,092,000	Nov. 14-Apr. 27	Released in Hatchery and Trap creeks ponds; 4,077,000 retained in troughs before release into ponds.

Table 19 — Continued

Brood Year	Total Release	Date of Release	Comments
1916	15,003,000	Nov. 27-April 27	Released in Hatchery and Trap creeks ponds.
1917	12,990,000	...	...
1918	19,852,000	Nov. 26-Jan. 31	Released in ponds
1919	11,357,000	Dec. 20-April 13	Released in ponds.
1920	18,913,000	Nov. 19-Mar. 21	Released in ponds.
1921	9,647,000	Dec. 3-April 7	Released in ponds.
1922	6,007,000	Dec. 7-April 12	Released in ponds.
1923	17,234,000	Nov. 22-Mar. 20	Released in ponds.
1924	19,430,000	Nov. 10-Jan. 30	Released in ponds.
1925	19,345,000	Nov. 20-Jan. 6	Released in ponds
1926	19,340,000	Nov. 4-Jan 14	Released in ponds.
1927	19,360,000	Dec.-May 8	Released in ponds.
1928	19,340,000	Nov. 7-Jan. 31	Released in ponds.
1929	11,285,000	Nov. 23-Mar. 31	Released in ponds.
1930	20,266,000	Nov. 27-Mar. 14	Released in ponds.
1931	19,403,000	Nov. 11-Mar. 21	Released in ponds.
1932	22,173,950	Nov. 15-Dec. 14 April 5-May 7	Released in ponds.
1933	20,030,000	Nov. 18-Dec. 30 Feb. 1-Mar. 31	Released in ponds.
1934	9,860,000	...	Released in ponds.
1935	32,531,000	Nov. 21-unknown	Released in ponds.

had the anal fin removed, and five had a section cut out of the middle of the anal fin. No known marks had been applied at Quadra and the hatchery had been closed in 1905, 1906, and 1907. No such fin mark had been applied and reported by other Alaskan hatcheries.

One tagging study in 1926 by Willis H. Rich and Arnie Suomela lends some credence to the success of the Quadra hatchery program.[65] Experiments were conducted at Cape Fox where 751 sockeye salmon were captured, tagged and released. Three hundred and eight were recatured and reported. The largest number of tags returned (49 tags) were from fish harvested at Boca de Quadra. The second largest return was to the Quadra hatchery (44 tags). The other returns were from the Nass River, Canada (35 tags), Tree Point (31 tags), Slate Island (22 tags) and Kah Shakes (16 tags). Tree Point and Slate Island were trap locations and thus intercepted fish. The Quadra hatchery logbook reports that actually 48 tagged fish were captured during the spawning season at the hatchery. Rich and Suomela made this observation:

> "By far the greatest number of these fish went north and were taken in Revillagigedo Channel, Boca de Quadra and Behm Canal. It is especially interesting to note the very large numbers taken in Boca de Quadra and at the Boca de Quadra hatchery. It is evident that this is one of the important producing areas for the red salmon runs that strike in at Cape Fox early in the season, and from the fact that so many of the tags were taken at the hatchery it seems possible that the hatchery operations are responsible for this unusual productiveness. In support of this we have the fact that runs to the hatchery have been increasing for a number of years."[66]

A look at the early records of Sockeye Creek, the outlet stream to Hugh Smith Lake, reveal that it had been considered one of the best sockeye streams in Southeast Alaska since commercial fishing began. In 1895, 97,000 sockeye salmon were taken from around its mouth and approaches; 137,000 were obtained in 1896; 65,000 in 1897. The stream continued to be heavily fished.

When Wisner made his examination in 1903, he found fishermen purse seining in the mouth of Sockeye Creek. That same year 17 purse seines were used by three companies. The hatchery reported there were no fish in the lake for use by the hatchery, and only 600,000 eggs were taken. The Quadra superintendent wrote that the cause of the failure was overfishing off the mouth of Sockeye Creek. He reported "There were through the whole of the season several cables across the mouth of the river and several parties had their camping and fishing scows, fishing and other boats, anchored there all the time, which prevented the fish from going up the stream and I am told on several occasions they fished beyond the limit line."[67] The limit line at that time was 100 yards outside the creek mouth.

Sockeye Creek is another of those pockets conducive to successful entrapment of entering fish. Its entrance, formed by precipitous rocky walls, is only 50 yards wide and opens into a small bay. Fishing was legal outside the "immediate approach" to the stream, but is is unclear whether the inner bay was considered "immediate approach."

No hatchery operations were carried on during 1905 through 1907, and harvest records are not available for that time. There were sufficient spawners in 1907 that six million eggs were taken for the Fortmann hatchery. Three million eggs were secured when the program revived at Quadra in 1908. That year the fishermen in the vicinity agreed that no fishing was allowed inside a line drawn from the old saltry site just outside the mouth of Sockeye Creek to the opposite headland. This saltry, built around 1890, was located on the northern shore, but it is hard to determine if it was inside or outside the inner bay. In essence, it would appear that fishing continued in a very constricted area.

The first year that fishing was not pursued vigorously off the mouth of Sockeye Creek was in 1912, but it was not until January 1, 1916 that the creek was closed 500 yards from its mouth. On January 1 1925, the locality was further protected within one mile of the mouth of the creek.

The sockeye run at Buschmann Creek was again utilized for hatchery purposes in the 1950s for the Territory's lake stocking program in the Ketchikan area. Many fish were removed from the system at that time. After statehood, the Alaska Department of Fish and Game maintained a weir on Sockeye creek for four years and records show the runs fluctuated greatly: 1968 - 1,617; 1969 - 16,429; 1970 - 8,755; 1971 - 22,096.[68] The run in 1981 was estimated to be round 11,000 sockeye salmon.[69]

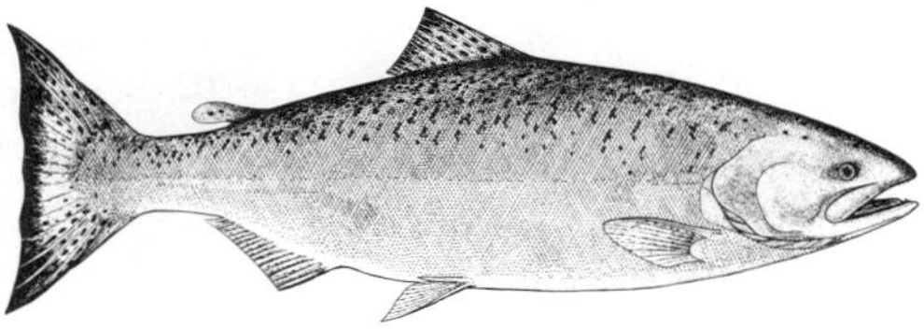

Notes

1. John J. Coyle, "Report of John J. Coyle, Assistant Agent for Protection of Salmon Fisheries in Alaska, 1903," Reports and Related Records, Division of Alaska Fisheries, RG 22, NA.

2. Ibid.

3. Ibid.

4. Ibid.

5. J.N. Wisner, "Alaska Reports, 1903," File "Hatchery Inspection Report, 1903-1910," General Records, Division of Alaska Fisheries, RG 22, NA.

6. Coyle, "Report, 1903."
The Mining Journal [Ketchikan], August 9, 1902.

7. Howard M. Kutchin, *Report on the Salmon Fisheries of Alaska-1902* (Washington: Government Printing Office, 1903), p. 29
Howard M. Kutchin, *Report on the Salmon Fisheries of Alaska-1904* (Washington: Government Printing Office, 1905).

8. Kutchin, *Report on the Salmon Fisheries of Alaska-1903* (Washington: Government Printing Office, 1904), p. 29.

9. Kutchin, *Report on the Salmon Fisheries of Alaska, 1904.*

10. Letter from H.J. Larrison to the Secretary of the Treasury, dated Summer, 1902. "Letters Received relating to Salmon Fisheries, 1/4/02-7/3/03," Alaska File of the Secretary of the Treasury, 1868-1903, Microfilm publication 720, Roll 21, National Archives.

11. Kutchin, *Report on the Salmon Fisheries of Alaska-1904.*

12. John N. Cobb, "Alaska Diary-1904, part 2", (John N. Cobb Collection, University of Washington Archives, Box 4, 11107 C).

13. Jordan Lake was named for David Starr Jordan, a leading ichthyologist at Stanford University. The lake was also known as Kutchin Lake in 1902 (for a U.S. Treasury agent in charge of Alaskan fisheries).

14. Patching Lake was named for Fred Patching, the Fortmann hatchery superintendent, who came to Alaska and was trained in fish culture by John C. Callbreath at McHenry Inlet.

15. Chamberlain Lake was named for Fred M. Chamberlain, a naturalist aboard the U.S. Fish Commission steamer *Albatross* and later a Bureau of Fisheries Alaskan agent. Chamberlain studied the salmon in the Naha River system from 1903-1913.

16. "Naha Salmon Hatchery," *The Mining Journal* [Ketchikan], September 27, 1902, p. 3.

17. *Pacific Fisherman,* January 1905, Vol. III, No 1, p. 46.

18. Millard C. Marsh and John N. Cobb, *The Fisheries of Alaska in 1907,* Bureau of Fisheries Document 623 (Washington: Government Printing Office, 1909), p. 27.

19. F.M. Chamberlain, "Report of the Work in Bristol Bay and Elsewhere in Alaska, 1913," Reports and Related Records, Division of Alaska Fisheries, RG 22, NA.

20. Letter from acting Commissioner of Fisheries to Alaska Packers Association dated July 13, 1916, File "Alaska Packers Association-Closure of Karluk, 1916," Reports and Related Records, Division of Alaska Fisheries, RG 22, NA.

21. Chamberlain, "Report of the Work in Bristol Bay and Elsewhere in Alaska, 1913."

22. Dennis Winn, "Season's Report, Southeast Alaska, 1925, " Reports of the Bureau of Fisheries Agents, Division of Alaska Fisheries, RG 22, NA.

23. Wisner, "Alaska Report, 1903."

24. *Pacific Fisherman,* May 1906, Vol. V, No 5, p. 11.

25. Willis H. Rich, "Notes from the Diaries of Willis H. Rich, 1922-1933," (National Marine Fisheries Service, Auke Bay Fisheries Laboratory library: Auke Bay, Alaska) September 17, 1923.

26. F.M. Chamberlain, *Some Observations on Salmon and Trout in Alaska,* Bureau of Fisheries Document No. 627 (Washington: Government Printing Office, 1907), p. 22.

27. Statement of Don Siedelman, Sport Fish Division, Ketchikan District, Alaska Department of Fish and Game, September 25, 1979.

28. Chamberlain, "Report of the Work in Bristol Bay and Elsewhere in Alaska, 1913."

29. John N. Cobb, "Alaska Diary, 1910," (John N. Cobb Collection, University of Washington Archives), June 22, 1910.

Millard C. Marsh and John N. Cobb, *The Fisheries of Alaska in 1910,* Bureau of Fisheries Document No. 746 (Washington: Government Printing Office, 1911), p. 11-12.

30. Statement of John Olafson in personal interview, Ketchikan, Alaska, March 18, 1975.

31. Chamberlain, "Report of the Work in Bristol Bay and Elsewhere in Alaska, 1913."

32. Chamberlain, *Some Observations on Salmon and Trout in Alaska.*

33. Fred M. Chamberlain and Ward T. Bower, "Fishery Industries," *Fishery and Fur Industries of Alaska in 1912,* Bureau of Fisheries Document No. 780 (Washington: Government Printing Office, 1913), p. 35.

34. Ward did not complete a report on these Alaskan studies, much to the distress of the Bureau of Fisheries who continued for a number of years to ask Ward for the results. Ward's papers at the University of Illinois Archives contain none of his Southeast Alaskan studies and only a few of work he conducted in the Copper Center-Cordova area in the summer of 1919. He has no papers at the University of Nebraska Archives.

35. Chamberlain, "Report of the Work in Bristol Bay and Elsewhere in Alaska, 1913."

36. George Embody, "Principles of Pond Fertilization," *Transactions of the American Fisheries Society,* 1928, Vol. IVIII, pp. 19-22.

37. Ward T. Bower, "Fish Culture in Alaska," *Alaska Fisheries and Fur Industries in 1911,* Bureau of Fisheries Document No. 766 (Washington: Government Printing Office, 1912), p. 73.

38. Barton Warren Evermann, "General Administrative Report," *Alaska Fisheries and Fur Industries in 1911,* Bureau of Fisheries Document No. 766 (Washington: Government Printing Office, 1912), p. 9.

39. Chamberlain and Bower, p. 33.

40. Ibid, p. 36.

41. Ward T. Bower and Henry D. Aller, *Alaska Fisheries and Fur Industries in 1915,* Bureau of Fisheries Document No. 834 (Washington: Government Printing Office, 1917), p.25.

42. Jefferson Moser, *Alaska Salmon and Salmon Fisheries* (Washington: Government Printing Office, 1899), p. 95.

43. Ibid, p. 96.

44. Chamberlain, *Some Observations of Salmon and Trout in Alaska,* p. 24.

45. Coyle, "Report, 1903".

46. U.S. Department of Commerce and Labor, Bureau of Fisheries, "Re: Prohibition of Commercial Fishing in Certain Streams and Rivers in Southeastern Alaska," Seattle, October 18, 1912.

47. Ibid.

48. Ward T. Bower, "*Alaska Fishery and Fur-Seal Industries in 1921,* Bureau of Fisheries Document No. 923 (Washington: Government Printing Office, 1922), p.10.

49. Ibid.

50. Letter from a citizen to Scott C. Bone, Governor of Alaska, dated September 26, 1922. Records of the Office of the Territorial Governor File 29-3-1922, Alaska State Archives, Juneau, Alaska.

51. Letter from Dennis Winn to Commissioner of Fisheries Henry O'Malley dated November 17, 1926. File "Alaska Packers Association-Fortmann Hatchery, 1927," General Records, Division of Alaska Fisheries, RG 22, NA.

52. Letter from Commissioner of Fisheries Henry O'Malley to William Timson, APA president, dated February 5, 1927, File "Alaska Packers Association-Fortmann Hatchery, 1927," General Records, Division of Alaska Fisheries, RG 22, NA.

53. Letter from William Timson to Henry O'Malley, dated February 11, 1927, File "Alaska Packers Association-Fortmann Hatchery, 1927," General Records, Division of Alaska Fisheries, RG 22, NA.

54. Ibid.

55. Cobb Creek was also called Old Hatchery Creek until 1910 when John N. Cobb, the assistant agent for Alaska Fisheries named it Nagel River after a federal official who visited Alaska in 1909. At some time after that it became known as Cobb Creek for John N. Cobb.

56. Wisner, "Alaska Reports, 1903,"

57. Buschmann Creek was named for Peter Buschmann, superintendent of the corporation which first operated the Quadra hatchery. Hugh Smith Lake was named for a Commissioner of Fisheries.

58. *Quadra Hatchery Logbooks, 1905-1935,* (Alaska Historical Library, Juneau).

59. Statement of Mark Meyer in personal interview, Washougal, Washington July 31, 1979.

60. Ibid.

61. *Quadra Hatchery Logbooks, 1905-1935.*

62. Meyer, July 31, 1979.

63. Letter from Commissioner of Fisheries Henry O'Malley to P.H. McCue, Northwestern Fisheries Company, dated February 5, 1927, File "Northwestern Fisheries Inc., 1917-1938," General Records, Division of Alaska Fisheries, RG 22, NA.

64. John N. Cobb, "Report of Inspection of Alaska Fish Culture Stations, Season of 1910," Reports and Related Records, Division of Alaska Fisheries, RG 22, NA.

65. Willis H. Rich and Arnie J. Suomela, "Salmon Tagging Experiments in Alaska," *Bulletin of the Bureau of Fisheries, Vol. XLIII, Part II, 1927* (Washington: Government Printing Office, 1927), pp. 72-78.

66. Ibid, p. 77.

67. Howard M. Kutchin, *Report of the Salmon Fisheries of Alaska-1903.*

68. Alaska Department of Fish and Game Weir Statistics, (Ketchikan District Office: Ketchikan).

69. Statement of Walt Larrick, Operations Manager (Southern Southeast Regional Aquaculture Association, Ketchikan.)

V. The Federal Government Hatcheries

YES BAY HATCHERY

Congress took the first positive step toward federal participation in the Alaskan hatchery program when it passed a sundry civil bill on March 3, 1905. This bill appropriated $50,000 for construction and operation of one or more salmon hatcheries in Alaska by the Bureau of Fisheries. The Bureau immediately began selection of a site with the view that the hatchery must be operational by the time spawn could be taken in the fall. H.C. Fassett, a Bureau employee familiar with Alaska, recommended three sites: 1) the lower end of the Wood River in Bristol Bay, 2) Afognak Island and 3) Yes Bay in Southeast.[1] Afognak Island had already been set aside by Presidential Proclamation as a fish culture reserve, and Yes Bay had been temporarily reserved for the same purpose.

Fassett recommended Yes Bay as the site of the Bureau's fish hatchery.[2] His site recommendation was reinforced by Barton Evermann, assistant in charge of the Division of Scientific Inquiry. Evermann stated "It is quite certain that it would be much easier to get a hatchery in operation this year at Yes Bay than at Afognak Island. Personally I consider it quite doubtful if a station could be built and made ready for operation at Afognak this year."[3]

The final decision rested with the Secretary of Commerce and Labor, under whose supervision the Commissioner of Fisheries operated. The secretary approved the proposed hatchery site at Yes Bay and had the solicitor of his department advise the President of the United States to permanently reserve the land. This Presidential Proclamation, made February 1, 1906, set aside portions of Yes Bay as a hatchery reserve. The reserve was described as both land and water area, including the lake and its catchment basin, its outlet and portions of the shoreline in Yes Bay and Back Bay.[4]

The selection of Yes Bay was based on several factors. First was its proximity to established transportation routes, a major advantage in 1905. Yes Bay was only 45 miles from Ketchikan which was one of the principal ports of Southeast Alaska. Steamers to and from Puget Sound arrived and departed daily for half the year and two or three times a week during the rest of the year. An established run by a mailboat also existed between Ketchikan and Yes Bay. Another factor was that the city would provide a temporary labor force and a sawmill had all necessary grades of lumber in stock. These factors lowered costs and saved time.

Yes Bay itself had many advantages. The bay is a narrow arm of the sea leading into the mainland of Alaska (Map 10). The Yes River, flowing into the bay, has a large lake as its source, first called Yes Lake and later McDonald Lake.[5] The major tributary at its head is known as Walker Creek, but was called the Yes River and Hatchery Creek during hatchery operation. This creek, with a falls, seemed ideal for a water and power supply. It was, in addition, the natural spawning ground for a large run of sockeye salmon.

Fassett foresaw some drawbacks. One was the necessity for construction of a tramway from salt water to the lake, a distance of less than a mile at one hundred feet of elevation. He also anticipated that Alaska Packers Association (APA) might complain if fishing in Yes Bay, upon which they relied for a large portion of their sockeye catch, was closed to them.

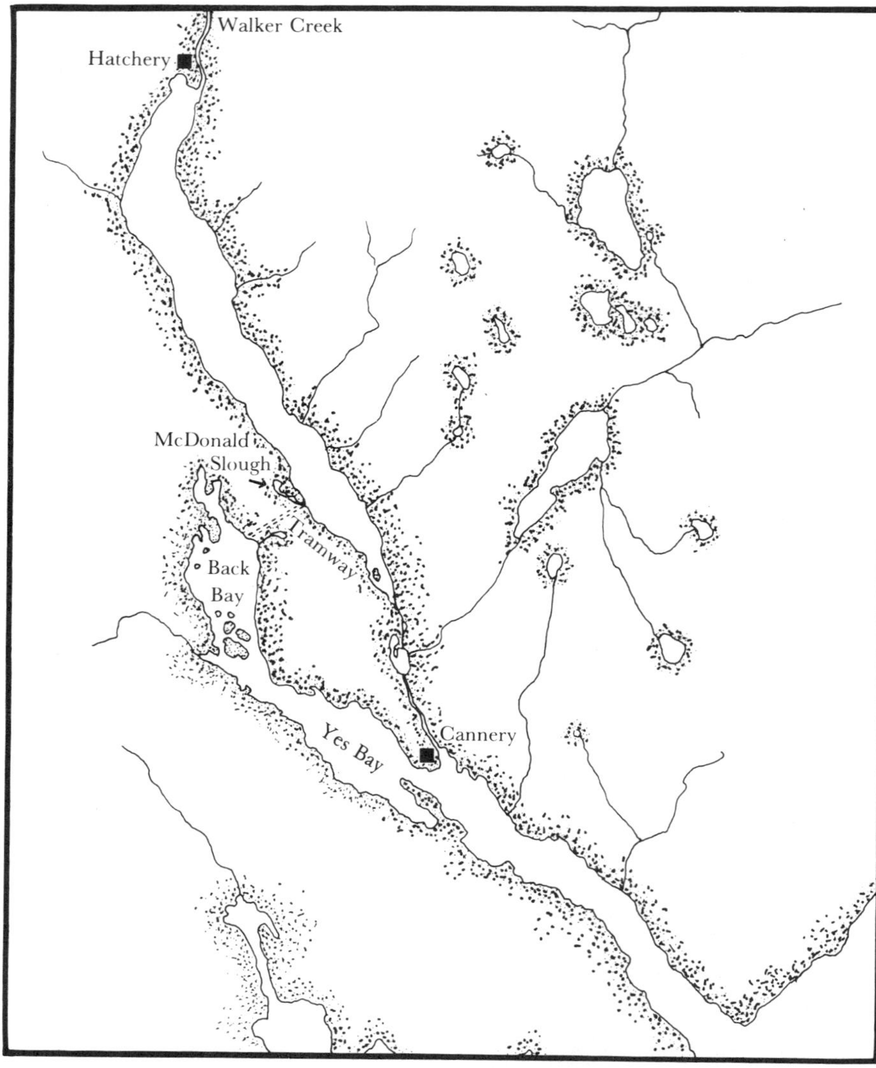

Map 10. Yes Bay and McDonald Lake

This company did not operate the cannery which was located at the mouth of the Yes River. This cannery was first operated in 1889 and became a part of the Pacific Packing and Navigation Company consolidation, a firm which failed. Yes Bay was thus opened to fishing by another company: in this case APA. At about the time the hatchery site was selected the bankrupt concern was purchased by Northwestern Fisheries Company, a company which operated hatcheries in Alaska.

The purchase prompted Fassett to include a warning to the Commissioner that Northwestern Fisheries Company might feel that they had vested rights to a hatchery site on the lake selected under the 1900 act. This proved to be needless precaution as the company sold the Yes Bay cannery to Charles Burkhardt and Company, who operated it under various names and who showed no interest in operating a hatchery.

Once the Yes Bay location was selected, Claudis Wallich was appointed to supervise the construction. Wallich, superintendent of the Bureau's Clackamas, Oregon hatchery and a former contractor and builder, was accomplished in fish culture and had published a paper which outlined a system of recording egg development using temperature units, thus creating a more efficient terminology for expression of rate of growth.[6]

Wallich and F.M. Chamberlain, at that time a naturalist from the *Albatross,* sailed for Alaska by commercial steamer on May 18, 1905, a little over ten weeks after the appropriation was made. Chamberlain was assigned to make a biological survey of the Yes Bay region. It was the first time either man had been to the site selected for the hatchery. After the pair arrived at Yes Bay, it was necessary to carry a rowboat to the lake and row four and a half miles to the head of the lake where they pitched camp.

They looked for a satisfactory water source for three days, but no springs could be located. Wallich and Chamberlain could not find anyone who had been to the lake in the winter so no information was available as to whether any stream remained open. Water temperatures were tested in several streams and with slight variations all were found to be the same as the inlet stream, Walker Creek, i.e., 52°F. Wallich finally recommended the water from this stream be used in the hatchery, saying "We can scarcely see why the water in which fish naturally deposit their spawn should be unfavorable for hatchery purposes, yet this seems to be an Alaskan prejudice."[7] Wallich did not question the water's purity or quantity, but was unsure of its reliability during the winter.

As the actual site for the hatchery building, the men chose a location one-fourth mile above the stream mouth close to the spawning grounds, and Wallich told the Commissioner:

> "The level, low lying ground, while evidently not subject to serious overflow, presents very favorable ground for ponds which may be

> connected with laterals directly to the river for water supply or may receive the waste water from the hatchery. Fry could be handled at a minimum cost and left to distribute themselves in the most favorable water the river and lake affords."[8]

After this intitial investigation, Wallich and Chamberlain made necessary preparations for construction of a hatchery at this remote site and returned with a crew and the *Albatross* several weeks later. The crew lived in quarters provided by the cannery at first, but finally moved to improvised quarters at the hatchery site in August.

The first project was a transportation system. The navigator of the *Albatross* made a number of sights for laying out a one-half mile long tramway from Back Bay to McDonald Lake. A gasoline launch and sufficient materials to build a 12 foot by 24 foot scow to haul supplies were taken over the completed tramway. This system of tramway and launches with barges was upgraded, increased, and maintained during the life of the station.

After completion of the transportation system, the water system was begun at the hatchery site. A trail 10 feet by 4,250 feet was slashed through the brush to the pool below the falls. The navigator again helped by laying out the flume's path. The wooden plank flume was four feet wide, one foot deep, and 1,200 feet long with a capacity of 4,000 to 5,000 gallons per minute. It was constructed through the creek bottom and along the edge of the bank protected by riprap. When the flume was nearly completed in September, Wallich took 7,440,000 eggs from a large run of sockeye salmon. Hatchery baskets, which had been constructed by the crew, were filled with eggs and placed in the upper part of the flume to incubate. These eggs were moved to the hatchery later in the winter when the building was completed.

The hatchery, designed by a Bureau of Fisheries architect and engineer, H. von Boyer, was typical of Bureau hatcheries in the United States (Fig. 40). Hatchery troughs were 16 feet long by 17 inches wide (outside measurements), of 1.5 inch lumber, and were placed in double rows of three troughs each with 1.5 inch fall between troughs. The water supply trough, 14 inches by 12 inches (inside measurements), was 1.5 inches above the first trough. Standard egg baskets were used in the troughs. Construction costs by April, 1906 were $24,130.09 and did not include the steam heating plant, which was estimated to cost $1,000.

Problems with the location of the new hatchery were immediately evident. Wallich's and Chamberlain's assessment that the level, low-lying land was not subject to serious overflow proved incorrect. The men moved to the hatchery building when all quarters and the messhouse had two feet of water over the floors in November, 1905. On the report to the Bureau's headquarters in Washington, D.C. an official wrote in the margin: "Good Engineering!"[9]

Figure 40. Yes Bay Hatchery Complex, circa 1929. *Courtesy A. Kemmerich.*

Another time the water was so high workers used a rowboat to get from quarters to the messhouse.

Floods not only caused inconveniences but also jeopardized hatchery operations. The flume raised during a flood in November 1907 and for two days there was danger of it washing away. Mud from the intake pool washed into the troughs but apparently did no damage. Thirty feet of flume washed away in 1917 but a pump was used until the flume was repaired.

Ice turned out to be the most serious problem. Severe winters occurred in 1905 and 1909, and the open flume froze solid both times, although no eggs were lost. Men worked day and night chopping a notch in the ice so that water could reach the hatchery. About 1,000 gallons per minute were normally needed for the hatchery. Only 500 gallons per minute were used in the winter when the water was near freezing. During 1905 it was thought that flow was further reduced to 300 gallons per minute.[10]

Attempts were made to improve the water supply system by Superintendent W.A. Hancock, who replaced Wallich in 1907. This task proved to be formidable. All changes and new construction in the facility had to be approved in Washington, D.C. and all funds came from Congressional appropriations. Against these odds, Hancock began his battle to replace the flume with a pipeline.

Hancock explained in one of his many letters that the greatest part of the flume was elevated from one to fifteen feet above ground and had a fall of three-eights of an inch in every 16 feet.[11] It was this portion above ground that froze nearly every winter. Someone in Washington, D.C. suggested in 1908 that Hancock place the flume underground. Hancock patiently explained the problems of such a task: digging, blasting, intense labor if put through on grade. There was no engineer at the hatchery, and the men were unfamiliar with running grade lines.

Hancock begged in 1907 for a pump which could be run by steam from the 30 horsepower boiler used to heat the hatchery in the winter. The river could be dammed and water pumped into the system. His suggestion was rejected and the men chopped ice 24 hours a day throughout cold spells.

Hancock renewed his request for a pipeline the next year. The architect, von Boyer, suggested that the flume be boxed in and packed with straw for insulation.[12] Hancock again patiently explained that the flume leaked, and it would saturate the straw which would freeze. Besides there was no straw at Yes Bay. The Washington officials and von Boyer were tired of the water problems and von Boyer wrote:

> "The practicability of the flume built only a little over a year ago at the cost of $2,500 should certainly have been verified before construction began, either by observation of how other private hatcheries in the vicinity were and are supplied, or through the advice of men competent to judge of this matter."[13]

He suggested a pump be purchased to run when the flume was inoperative.

Hancock finally received the pump which was installed and used during the winter of 1909 when the flume froze again, ice forming on the bottom and sides of the flume until it was filled. The pump ran until the men chopped the ice out of the flume.

Hancock was also instructed to build a new flume inside the old one, with drainage holes drilled in the bottom of the old flume. A four inch space between the flumes would be packed with spruce needles. Hancock protested this eliminated the ability to produce electrical power by reducing the head. Light was necessary, Hancock explained, for picking eggs during the winter months when daylight hours were short. Washington officials felt installation of an acetylene system was less expensive, and it had proved to be successful in other Bureau hatcheries. The flume modifications were made.

The acetylene system was installed in 1909 and used until 1914 when electricity finally was acquired at the station. All facilities were heated with wood, however, so the employees spent much time logging, cutting rounds, splitting and distributing wood to all buildings. About 200 cords of wood were used each season in later years.

Hancock did not win his battle with the Washington officials to replace the flume with a pipeline. The pipeline was finally installed during the reign of one of his successors when the station was rebuilt in 1920. Considerable construction and repair essential to proper maintenance of the station were undertaken. The condition of the water supply flume had been the cause of much worry. When it appeared the flume had reached the full extent of its endurance at the end of the fiscal year 1919, the annual hatchery appropriation was used for installation of the pipeline. This gave the hatchery a water supply with ample reserve for future expansion. An allowance was made for furnishing power for a lighting plant. The station was closed to fish culture work in 1920, and a new foundation placed under the hatchery, replacement hatchery troughs installed, and 6,000 feet of 16 inch by 18 inch wood stave pipe laid, 2,300 feet of it on trestles.

Fish culture work resumed at the Yes Bay hatchery in 1921. Unfortunately the pipeline traded one set of winter woes for another. Employees no longer cut ice out of the flume to assure a flow of water. Instead when anchor ice formed in the intake pool, the crew chopped ice, often throughout the night. This effort was necessary throughout the life of the station.

Although the site created problems from ice and floods, Wallich and Chamberlain were correct in their assessment of the proximity of the spawning grounds. Walker Creek had the major run of sockeye salmon in the lake. A rack or weir, over 200 feet long, was erected each July. Piling were driven and a permanent apron was embedded in the stream in 1920 and replaced in 1929. The wooden slatted rack pieces were numbered and so arranged that when taken out in the fall, they would be in regular order for use the following season. Despite every precaution, some years the rack washed out during high water.

Beach seines, 450 feet long, were used to capture the fish below the rack. The seine was strung from the rack, then a crew of 11 or 12 men, each clad in waders, pulled the seine around the fish. Alphonse Kemmerich, a former superintendent, remembers that six million eggs were spawned in 1932 from one seine haul.[14] Once captured, fish were tested for ripeness and green fish released. Standard spawning procedures adopted by the Bureau of Fisheries were used. The spent carcasses were thrown into the lake until 1914 when fish were salted in barrels for use as fish food.

Sockeye salmon were the major species from which eggs were taken. Superintendent Hancock (1907-1916) started taking sockeye eggs on September 1 to 3. Later superintendents usually started the second week in September. The spawning season depended upon many factors, but generally ran from four to six weeks. Factors contributing to the number of eggs taken

Table 20. Sockeye Spawn Taken, 1905-1932: Yes Bay Hatchery

Brood Year	Number of Eggs Taken	Spawning Period	Limiting Factors
1905	7,440,000	...	Incubation capacity.
1906	58,210,000	...	Incubation capacity.
1907	65,550,000	Sept. 3-Sept. 28	Incubation capacity.
1908	50,000,000	Sept. 2-Oct. 6	Flood conditions, fish escaped above rack.
1909	72,000,000	Sept. 2-Sept. 30	Incubation capacity.
1910	72,000,000	Sept. 1-Sept. 24	Incubation capacity; 1,500,000 eggs transferred.
1911	72,000,000	Sept. 1-Sept. 28	Incubation capacity; 2,000,000 eggs transferred.
1912	66,125,000	Sept. 1-Oct. 4	Insufficient fish available; 2,000,000 eggs transferred.
1913	49,050,000	Sept. 1-Oct. 2	Insufficient fish available; 2,000,000 eggs transferred.
1914	41,300,000	Sept. 3-Sept. 30	Insufficient fish available; 3,000,000 eggs transferred.
1915	72,000,000	Sept. 2-Sept. 29	Incubation capacity; 18,100,000 eggs transferred.
1916	58,000,000	Sept. 1-Oct. 2	Insufficient fish available; 2,000,000 eggs transferred.
1917	34,950,000	Sept. 6-Sept. 26	Insufficient fish available.
1918	47,300,000	Aug. 29-Oct. 2	4,500,000 eggs transferred.

were the capacity of the hatching baskets, the height and the velocity of the creek which dictated whether haul seining was possible, washouts of the rack, and the philosophy prevailing at that time on treatment of the fry. The factors are reflected in Table 20.

Once the eggs were fertilized, they were placed in baskets and incubation proceeded. Most eggs were run through the salt solution during the month of November to remove dead eggs.

Little specific incubation data have survived, but the eggs taken during the early part of the run (when the water was warmer) generally began to hatch in 80 days, while eggs taken near the end of the season (when the temperature was cooler) began to hatch in about 140 days. The mean water temperatures as recorded in 1913 were: September 51.7 °F; October 46.5 °F; November 38.5 °F; December 34.5 °F; January 32 °F; February 33.6 °F; and March 33.0 °F.[15]

Table 20 — Continued

Brood Year	Number of Eggs Taken	Spawning Period	Limiting Factors
1919	9,752,000	Sept. 11-Sept. 25	Delayed egg take and loss of rack on sixth day of spawning.
1920	...	...	Renovation of hatchery. No eggs taken.
1921	51,000,000	Aug. 29-Sept. 28	Flood conditions hampered capture of adults. 150,000 eyed eggs planted in nearby lakes.
1922	25,000,000	Sept. 5-Sept. 23	Rack washed out.
1923	25,500,000	Sept. 6-Sept. 21	Rack washed out.
1924	30,800,000	Sept. 8-Sept. 25	Fry capacity.
1925	39,680,000	Sept. 7-Oct. 6	Fry capacity; 8,645,760 eggs transferred.
1926	30,760,000	Sept. 6-Sept. 30	Flood conditions hampered adult capture; 6,958,890 eggs transferred.
1927	20,240,000	Sept. 6-Oct. 6	Insufficient fish available.
1928	24,036,000	Sept. 6-Oct. 1	2,493,000 eggs transferred.
1929	16,095,000	Sept. 3-Oct. 3	Insufficient fish available.
1930	27,469,000	Sept. 6-Sept. 25	3,055,000 eggs transferred.
1931	25,000,000	Aug. 29-Sept. 17	3,144,960 eggs transferred.
1932	25,500,000	Aug. 29-Sept. 13	3,010,650 eggs transferred.

Once the eggs began to hatch, it was the policy through brood year 1917 to release sac-fry into Walker Creek during December, January and February (Figs. 41 and 42). The egg capacity was greater than the fry capacity and once the eggs began to hatch, the sac-fry had to be thinned out. The troughs were refilled after the first hatched sac-fry were released. This continued until there was sufficient trough space for the remaining sac-fry which were then held for up to a month.

The stacked tray system described in Chapter II was adopted after 1921 and, thus, the same number of sac-fry as eggs could be held. No more sac-fry were planted after that time except in emergencies due to water supply problems. A summary of the fish released is shown in Table 21.

The first attempts to feed fry were experimental and designed to determine the best diets. Practically nothing was known about what fry would eat and

Figure 41. Transport of sac-fry to release site, Yes Bay hatchery,before 1914. *National Archives.*

Figure 42. Release of sac-fry in river, Yes Bay hatchery,prior to 1914. *National Archives.*

thrive upon. In the first experiment on the 1907 brood, five lots, totaling 25,000 fry, were fed different types of food: canned salmon, canned roe, canned milt, fresh liver and canned liver. Within a month all but 1,000 fish had died because the majority would not feed and then became diseased.

Better results occurred in 1908 when a smaller number of fry were fed on chopped steelhead. These were fed until October 14 and released into the lake. In 1909, 56,000 number one fingerlings were held in the hatchery and fed ground fresh fish, one pound fed daily per 18,880 fry. No record was found of how these fry survived.

The next year, 1910, fry were fed herring roe upon which they did badly, but recovered when fed ground salmon. Fish were again fed canned herring roe in 1912, but the fish became very thin and emaciated and losses were heavy. Herring roe having proven unsatisfactory, the diet was changed to raw, ground Dolly Varden and the fish began to grow. Some problems were encountered when holding the sockeye fry in fresh water. A salt bath was given in June, and it was necessary a month later to give the fry a vinegar bath. There is no report on the survival on this group of fry or when they were released into McDonald Lake, although they were reported to be two inches long on the first of September, 1912.[16]

Once feeding was past the experimental stage, diet for the fry depended upon what was available and consistent with previous experiments. Location of quantities of food was a constant problem. To help alleviate the problem, spawned salmon were mildcured each fall starting in 1914.

Salt salmon were cooked, pressed and grated into a meal in a meat chopper operated by power from the turbine. At first the fingerlings seemed to thrive on the food, but then growth arrested and heavy mortality ensued. There was some question whether this was due to the condition of the food or to the high temperature of the water during the period of feeding. The next year, 1916, salt salmon were freshened and ground, but not cooked, and better fry development was obtained. Salt salmon never proved a satisfactory food because the freshened flesh was generally very soft and had lost much of its food value.

Consequently, other food was constantly being sought. Gillnets placed at the mouth of Walker Creek every spring captured Dolly Varden and steelhead which were then ground raw and fed to the fry. Not only did this furnish food, but it controlled what were considered predators. Fresh salmon eggs were obtained some years from the Yes Bay cannery. The fry were started on these eggs and/or beef liver when possible.

Once space in the cold storage at New England Fish Company became available, sufficient fish food could be stored. A man was sent to Ketchikan each season from 1929 until 1932. Eggs from the salmon were frozen and stored

Table 21. Sockeye Released, 1905-1932: Yes Bay Hatchery

Brood Year	Number Released	Dates of Release	Distribution[a]
1905	6,638,500	May 23-June 5	Unknown.
1906	54,610,800	Dec. 21-May 31	Inlet stream as sac-fry (42,572,800); as fry (12,038,000).
1907	61,369,000	Dec. 2-Mar. 27	Upper inlet stream as sac-fry.
1908	48,653,000	Feb. 15-Oct. 14	Upper inlet stream as sac-fry (30,300,000); as fry (18,353,000); as #1 fingerlings (74,600); as #2 fingerlings (24,500).
1909	69,879,600	Jan. 6-May 28	As sac-fry in inlet stream (48,160,000); as #1fingerlings in lake (17,620,000).
1910	68,239,900	Dec. 23-June 30	As sac-fry in inlet stream (39,780,000); as fry (11,068,900); and as fry in lake (17,391,000).
1911	68,335,000	Dec. 8-June 29	As sac-fry in inlet stream (49,140,000) and as fry in lake (19,195,000).
1912	60,422,100	Dec. 16-June 19	As sac-fry in inlet stream (20,531,800).
1913	43,401,400	Feb. 4-July 26	Fed at least 6,334,000 fry in ponds. Unknown number of fingerlings released July 26.
1914	36,720,000	Jan. 9-April 19	5,000,000 fry released in ponds for feeding and released into lake from June 14-July 27.
1915	52,317,500	Dec. 21-July 18	As sac-fry and fry.
1916	51,175,500	Dec. 19-July 30	As sac-fry and fry. Some fry fed.
1917	32,539,200	Jan. 19-June 11	Unknown.
1918	35,329,700	Feb. 13-July 10	McDonald Slough used to feed some fry until Sept. 17 when fence washed out.
1919	9,387,000	June 1-July 10	3,000,000 fry released into McDonald Slough on July 1.
1920	4,025,000	May 11-May 12	As sac-fry (250,000) on Feb. 23. 2,000,000 as fry into McDonald Slough. 1,000,000 as fry along lake shore. All fish from eggs transferred from Litnik Lake, Afognak Island.

Table 21 — Continued

Brood Year	Number Released	Dates of Release	Distribution[a]
1921	47,640,000	April 10-July 1	As sac-fry along lake shore (8,000,000); as #1fingerlings along lake shore (6,000,000); as fry along lake shore and in inlet stream (7,000,000); in McDonald Slough as #1 fingerlings (2,000,000) May 11-Sept. 10.
1922	24,099,045	May 11-July 11	All fish fed starting May 9. Released as #1 fingerlings along lake shore May 21 (1,500,000); into McDonald Slough on May 11 and released July 11 (6,500,000).
1923	21,817,800	Mar. 13-May 20	As #1 fingerlings (5,700,000); as fry along lake shores (3,200,000).
1924	27,328,000	Dec. 23-Oct.	As sac-fry when water was too low (3,900,000); Remainder fed from May 1 as long as Oct.
1925	27,392,200	Feb. 18-July 11	Mild winter so fish placed in ponds after swim-up. Last fry put in ponds July 11 and fed until after August 8.
1926	20,467,000	April 2-Aug. 22	Fry placed in ponds and allowed to work into lake volitionally.
1927	18,748,715	April 11-unknown	Fry placed in ponds and released in July and Aug.
1928	20,440,000	May 28-June 15	As #1 and #2 fingerlings (18,000,000).
1929	14,332,000	June 20-Sept. 8	As #1 fingerlings (6,002,000); as #3 fingerlings (8,000).
1930	22,480,000	Mar. 20-July 16	All in ponds, feeding beginning Mar. 21, gates opened July 15 but feeding continued until food supply exhausted.
1931	18,980,000	May 3-Oct. 18	All in ponds, feeding beginning May 3, gates opened in Oct. but feeding continued until food supply exhausted.
1932	16,533,000	May 13-July 25	All fry released by order of Commissioner of Fisheries. Many fish diseased.

(a) numbers of fish distributed will not add to total release because station logbooks do not give specifics on all fish distributed.

until needed. Two hundred tons of fresh trap-caught chum salmon were procured in 1932 in Ketchikan and placed in cold storage for feed in 1932 and 1933. These were ground whole after removal of head, tail and fins.

After these first attempts to feed the fry, experiments on how and where fry would be fed were conducted. The hatchery superintendent constructed 160 new hatching troughs in 1912 which would permit the rearing of from 35 to 40 million fry to the feeding stage. The loading densities were six quarts of fry per trough. One load of fry was removed and released by July 2 because they were not doing well. The remainder were given a salt bath but their fate is unrecorded.

Fry ponds were first constructed at the Yes Bay hatchery in the spring of 1914. The three ponds were each 60 feet long, 12 feet wide, and 2 feet deep at the upper end with a sloping bottom, giving a depth of 2.5 feet at the lower end. Each was lined with heavy planks, and the bottoms and sides caulked to make them watertight. Located side by side in a series between the hatchery and the creek, the ponds derived their water supply from the hatchery overflow. Water flow was generally 35 gallons per minute, but 50 gallons per minute could be obtained by turning more water into the hatchery headbox. The ponds were stocked in 1914 with 211,400 fry, although the suggested stocking had been 5 million. The foreman keeping the logbook wrote on May 23, 1914: ". . . This sum would be far too great as born out by the fact that even the above sum appears too great." [17] Ponds were cleaned several times during the season, but the logbook does not give specifics on methods.

Some fry were fed in the hatching troughs in 1916. It was discovered that these young fish thrived better in these troughs than in the rearing ponds.[18] The following year all fry were held in troughs and the ponds were subsequently filled and became part of the lawn.

A change in superintendents in 1918 brought yet another innovation for feeding large numbers of fry, adapted from a system used in Washington hatcheries. An arm of McDonald Lake, locally known as McDonald Slough, was temporarily screened, gillnets set to remove predators, and fry placed in it during the spring of 1919. Fry received regular rations of salt salmon. Midseason high water took out the barrier fence across the mouth of the four acre slough and the fry escaped into the lake.

Piling were driven in 1920 to make the feeding enclosure permanent. Screens, about four feet deep, made the barricade. Three million fry from transplanted eggs from Litnik Lake, Afognak Island in 1920 and two million fry in 1921 were held in McDonald Slough. The growth of the fry in 1921 was excellent and within 11 weeks fry had grown to two inches. Seven weeks later fry were permitted to migrate into the main body of the lake.

Predators were removed each spring from McDonald Lake prior to release of fry. Gillnets set within the compound overnight on May 5, 1923, captured 41

Dolly Varden in one net and 60 Dolly Varden and cutthroat trout in another net. Two shots of TNT in the slough did not kill any more predators, but Kemmerich remembers the foreman using about ten bottles of explosives with the same results.[19]

Over 6.5 million fry were placed in the slough from May 11 to June 6, 1923. Holding the fry in a portion of the lake proved unsatisfactory because of the warm, dry summer. McDonald Slough was an arm of the lake with no source of fresh water. Water temperatures began to rise. Each day the hatchery personnel expected the hot spell to break. Finally the gates were opened on July 11. The water temperature was 69 °F on both the surface and the bottom. The loss of fry was terrific according to Kemmerich. Rearing of fry in McDonald Slough was not resumed.

Another method of holding fry began during the major reconstruction in 1920. A thousand foot space at the upper end of a branch of Walker Creek which ran in front of the hatchery was divided into numerous feeding ponds by placing semi-circular rock dams across the stream. A six foot gauge controlled the water. A wire trap placed at the lower end of the creek kept out Dolly Varden and cutthroat trout.

The protected area where the fry were fed, the edges of the grass, and numerous little indentations along the bank provided more natural conditions than previous ponds. The fingerlings reared in these ponds were released into the lakes as it became necessary to add more fry. The last fingerlings volitionally passed into the lake, but feeding continued at regular intervals as long as fish remained in the ponds. This method of holding fry for feeding continued until the closure of the station in 1933. The holding of fry in the hatching troughs continued over the years and the resultant fingerlings were planted along the shores of McDonald Lake, generally before the taking of spawn in September.

Other species of salmon, steelhead, and trout were handled at the Yes Bay hatchery (Table 22). A few coho eggs were incubated during the seasons of 1908 and 1909. After this time, the coho run was captured not for artificial propagation, but for fish food. The carcasses were skinned and then salted. This was a late run of coho salmon and toward the later years of the station's existence the run was ignored, because maintaining the rack was extremely difficult due to periods of high water. The chum salmon population, in contrast to the coho run, was very small. Only once, in 1930, were chum eggs incubated and fry released into the lake in February and March.

A strong pink salmon run existed and the station handled some of this species, mainly for export as eyed eggs (Table 22). Field stations were necessary other times when the Yes Bay hatchery personnel had a request for pink eggs for export. Eggs from Ketchikan Creek were eyed in a building in that city for several years. When small runs of pinks entered Yes River and Ketchikan Creek in 1914, a substation was set up at Mink Arm, one of the several arms of Boca

Table 22. Pink, Coho, and Chum Salmon, Steelhead and Eastern Brook Trout Production, 1906-1933: Yes Bay Hatchery

Brood Year	Species	Number of Eggs Taken	Number of Fish Released	Distribution
1906	steelhead	182,000	143,500	Released in McDonald Lake.
1908	coho	17,000	9,900	Released 990 fry in lake June 10, 1909. 7,000 impregnated with sockeye milt; all died.
1909	coho	5,000	...	Unknown
1910	pink	114,000	...	Unknown
1913	pink	5,275,000	4,500,000	Eggs taken at Ketchikan Creek except for 35,000 from inlet stream. All planted as sac-fry in inlet stream Jan. 2-9.
1915	pink	325,000		Unknown
1918	pink	1,365,000	930,000	406,000 eyed eggs transferred. Unknown distribution of remainder.
1921	pink	246,000	210,000	Inlet stream or lake as fry.
1927	Eastern brook	54,000	41,600	Eyed eggs transferred from stateside. Released fry in lakes near Juneau, Sitka and Petersburg.
1928	pink	2,300,000	...	1,851,550 eyed eggs transferred. Unknown distribution of remainder.
1929	steelhead	150,000	82,000	Released as #2 fingerlings in Yes River August 19, 1930.
	pink	2,650,000	...	1,021,000 eyed eggs transferred. Remainder released as fry in feeding ponds for volitional release March 5-April 8, 1930.

Table 22 — Continued

Brood Year	Species	Number of Eggs Taken	Number of Fish Released	Distribution
1930	pink	7,055,000	261,000	6,107,000 eyed eggs transferred. Remainder as sac-fry in inlet stream March 21, 1931.
	chum	100,000	40,000	Released Feb. 18, 1931 as fry in lake.
			45,000	Released March 21, 1931 as fry.
1931	pink	353,000	66,000	Released as fry in inlet stream Feb. 25, 1932.
			250,000	Released as fry on March 31, 1932.
	Eastern brook	128,000	116,000	Eyed eggs transferred possibly from Oregon Game Commission. Released as fry in various lakes near Juneau, Ketchikan, Wrangell, Petersburg, Red Bluff.
1932	pink	558,000	153,000	379,904 eyed eggs transferred. Remainder released as fry on March 23, 1933.
	Eastern brook	50,000	36,000	Eyed eggs transferred from stateside. Released as fry in Southeast Alaska.
1933	Eastern brook	50,000	...	Unknown distribution.

de Quadra. Long-distance shipping cases were filled with 250,000 eggs and transported by chartered power boat to an egg eyeing station in Ketchikan. The quality of the eggs proved to be fairly good. Out of the collection of 2,600,000 eggs, the loss in eyeing was only 600,000 eggs. Heretofore it had been considered almost impossible to move green eggs any distance without a heavy loss.[20]

Protests by Ketchikan citizens about removal of pink salmon from Ketchikan Creek necessitated abandonment of operations there in 1916. About 325,000 pink eggs already taken were deposited in the gravel of the creek near the egg collection site. Searching for a new site, the crews selected Smeaton Bay, but only 29,000 pink eggs resulted and these were planted in nearby spawning beds. The export of pink eggs was not initiated again until 1928 and 1929. Eggs were collected those two years in Walker Creek and in a stream at the head of Yes Bay. Although most pink eggs were exported, some attempts were made to incubate and release pink salmon from the hatchery. Table 22 shows the number and the age of fry when released.

A trout program was initiated in the late 1920s. Steelhead eggs were incubated twice, but this species received little emphasis. A few ripe fish were caught in 1929 and subsequently spawned. Generally, as with cohos, the steelhead were ground for fish food. Eastern brook trout eggs were transferred to Yes Bay hatchery during four different years: 1927, 1931, 1932, and 1933. The Bureau introduced this species into Alaskan waters at the request of sportsmen and U.S. Forest Service personnel. The eyed eggs were incubated and the resultant fry fed. Two-inch fry were flown by float plane to lakes in various parts of Southeast Alaska. The Bureau distributed some fish and the U.S. Forest Service made other releases. Some fingerlings were also given to individuals for release in favorite lakes. While no releases of eastern brook trout in McDonald Lake were made, 50,000 fry were on hand when the station closed, and there is no mention of their distribution. It is probable that these fry were released into McDonald Lake.

Little evaluation of the success of the Yes Bay hatchery program occurred. Outmigrant checks appear to have been more from curiosity than from a scientifically designed plan. A net was set in the lower end of the lake on May 22, 1909. The logbook records that 50 sockeye fry (1.5 to 2 inches long), 40 coho smolts (3 to 6 inches long), 12 Dolly Varden (6 to 8 inches long), and a few rainbow trout and bullheads were caught overnight.[21]

The only published account concerned observations made during May 1913.[22] Just above tidewater, at a point where the Yes River is about 20 feet wide, a small net was made fast ashore at one end and to a large rock only three feet out from the same shore at the other end. Fourteen hours later this small section of net contained 1,310 yearling sockeye salmon, averaging three inches in length. The station logbook records that the fyke net captured 960 dead and

350 live sockeye yearlings and 350 fry in the net. The mortality can be attributed to the pressure of that great volume of water flowing through the fyke net.[23]

A net, according to the published account, was set again near the foot of the lake where the Yes River is about 50 feet wide.[24] This net, with the leadline about a foot from the river bottom, covered ten feet of the stream. The next morning 575 yearling sockeye salmon were taken from the net. The net was again set in place, but the high water hampered its effectiveness and only 50 yearling sockeye were caught. All fish were reported to be in excellent condition. The fisheries agents' annual report gives the following analysis of the outmigrant captures:

> "The fact that a small piece of net only three feet long, set in a stream 20 feet wide, will catch 1,310 sockeye yearlings in one night would seem to be very good evidence that the aggregate number of young salmon thus moving to the sea during the entire period of the spring migration is of highly satisfactory proportions. That the Yes Bay hatchery is a potent factor in this result almost goes without saying, as practically all of the sockeyes passing into Yes Lake are spawned at the hatchery."[25]

One other attempt was made to capture outmigrants. A fyke net was set at the lower end of the lake in 1917 to obtain specimens of fry and fingerlings for Dr. Charles H. Gilbert, who conducted research for the Bureau of Fisheries. The first sampling from May 2-10 involved a faulty fyke net and only 14 specimens were obtained. After repairs, it was set again on May 20 and between that date and June 16, 95 specimens were taken, the largest numbers on May 20 (20 specimens) and June 1 (50 specimens). It is impossible to determine from existing records if the nets used in 1913 and 1917 were the same type and size and if they were set in the same place. Comparison of the two year's data is not valid.

Visual assessments of the fingerlings leaving the lake were made by hatchery personnel. During the five years that Kemmerich was foreman and later superintendent of the hatchery, he observed that the sockeye fingerling migration occurred in June and July. At that time, generally on calm evenings, schools of possibly 10,000 to 50,000 fish could be observed on the lake surface moving down the lake.[26]

There was little attempt to evaluate the adult returns. Evidence of only one marking program was found.[27] This occurred in the fall of 1909 on 1908 brood fingerlings which had been fed. One hundred eighty fish divided into three lots of 60 fish each were marked with fin clips (dorsal and adipose clip, half dorsal and adipose clip, and ventral and adipose clip). No information has survived on adult captures. This does not mean there was no attempt to find adults with marks in the following years, but rather that the hatchery logbooks do not contain such information.

Historically the sockeye runs to McDonald Lake were considered extremely valuable commercially and salmon processing began in Yes Bay twenty years before the hatchery was constructed. A saltry was built there in 1886 and was converted to a cannery in 1889. Moser reported the following sockeye harvests from the Yes River which he compiled from company records: 1898 — 44,271; 1899 — 69,000; and 1900 — 80,000.[28] Moser estimated the Yes Bay system could produce 50,000 sockeye for harvest per year.[29]

During this time a pile trap was used to capture salmon for the cannery. For at least twelve years, from 1890 to 1902, this trap fished the mouth off the Yes River (Fig. 43). Fisheries Agent Howard Kutchin believed "every pile of it was illegal" as it occupied three-fourths of the width of the Yes River at high water.[30] However, through technicalities, the trap was legal and remained in place until 1903. Kutchin said of the trap:

> "There seems to be no marked falling off of fish. The case presents another of those apparently inexplicable puzzles that are constantly coming up and which seems to confound those who have made a study of salmon breeding."[31]

Catch records show that during the period of 1893 to 1900 with the trap in place, the highest catch of sockeye occurred in 1900. That year 80,000 sockeye

Figure 43. Fish Trap at Mouth of Yes River, 1890-1902. *U.S. Fish Commission.*

were reported to have been harvested (Table 23). The data shown in this table after 1906 are questionable. Records compiled by canneries were sent to two Bureau of Fisheries employees, Willis H. Rich and Edward M. Ball, who put together tables showing locations fish were caught.[32] Yes Bay was generally closed to fishing from 1906 to 1926, so it is a puzzle how the cannery recorded some of the larger catches which are shown on the table. For example, in 1910 when 139,143 sockeyes are listed, the hatchery was filled on September 24 with 72 million eggs. Fishing in the bay had been closed to commercial fishermen on August 17. This would mean that the run was over 200,000 fish, which was unprecedented. The smaller numbers of sockeye listed after 1912 could represent fish which were caught before the bay closed each year.

The annual closure of Yes Bay was established by the Presidential Proclamation dated February 1, 1906 which set aside the site for a hatchery reserve. It stated that fishing was permitted only when, in the opinion of the hatchery

Table 23. Commercial Catches of Salmon Attributed to Yes Bay, 1893-1926[a]

Year	Coho	Chum	Pink	King	Sockeye
1893	...	...	...	...	26,292
1894	...	...	...	...	21,541
1895	...	...	...	...	42,007
1896	...	...	...	...	46,706
1897	9,511	5,862	185,608	...	60,900
1898	6,413	...	45,000	...	44,271
1899	6,300	500	75,000	...	69,000
1900	7,700	2,250	60,000	...	80,000
1904	...	...	16,768	...	19,679
1906	1,500	...	...	...	18,000
1907	...	352	95,579	...	31,599
1908	2,163	871	39,451	...	47,233
1909	45	537	56,124	...	85,598
1910	1,366	6,430	155,991	...	139,143
1911	33	5,850	155,974	61	81,750
1912	...	277	8,630	...	1,014
1913	22	514	42,991	...	337
1914	55	547	479	...	17
1918	1	177	5,237	...	11
1919	...	256	921	...	283
1922	...	...	...	...	2,526
1923	2	20	906	...	72
1924	2	95	4,227	...	589
1925	50	1,506	8,813	...	61
1926	175	154	506	6	205

(a) See text for validity of statistics.

Compiled from Rich, Willis H. and Edward M. Ball, 1935. "Statistical Review of the Alaska Salmon Fisheries. Part IV: Southeastern Alaska," in *Bulletin of the United States Bureau of Fisheries, Vol. XLVII* (Washington: Government Printing Office.)

superintendent, salmon had entered Yes Bay in sufficient numbers to supply the maximum number of eggs required. The superintendent could stipulate the kinds and amounts of gear to be used, places they could be employed, and times fishing could be done. During the run of salmon each season, a deputy from the hatchery stationed at the bay saw that regulations were observed.

Thus, commercial fishing took place in Yes Bay only under the supervision of the hatchery superintendent. When in his judgment the run exceeded the numbers required for hatchery purposes, commercial fishing was allowed. During the first few seasons not enough fish entered the lake, and a shortage of eggs occurred because the number of fish entering the lake could only be estimated by the apparent abundance in the bay and the stream. It became the practice to keep the bay closed.

Restriction of commercial fishing, however, did not extend past the narrow mouth of Yes Bay. The companies, unable to fish in the bay, placed two floating traps, one on either side of the mouth. These two traps were not the only ones which could intercept sockeyes bound for Yes Bay, but were undoubtedly most effective for fishing sockeye salmon destined for the Yes River (Map 11).

The traps appear to have been first fished in 1911 or 1912. The superintendent of the Yes Bay hatchery complained, but his supervisors were unable or unwilling to do anything about the traps.[33] The traps were legally placed on either side of the bay's mouth and the full width of the bay was not obstructed. Both traps closed during a period from 6 p.m. Saturday to 6 a.m. Monday each week.

Government officials agreed that the traps were catching considerable numbers of fish. The only concession the superintendent received was that Fisheries Agent Ward T. Bower received verbal assurance from the trap owner that the traps would not fish when the run of sockeye became heavy. "It remains to be seen if he lives up to his word," Bower wrote.[34] No further correspondence on the subject was found, but on June 24, 1924 Yes Bay was further protected by the prohibition of fishing 1,000 yards from Bluff Point and Syble Point at the entrance of the bay, thus, eliminating the two traps. Others in the vicinity continued to operate.

The Yes Bay hatchery operated eight more years after the traps were removed. The order for the closure of the hatchery arrived on July 23, 1933, about six weeks after a visit from a newly appointed Commissioner of Fisheries. All fry were liberated immediately, despite the fact that a serious epidemic of disease was in progress (see Chapter II).

The station was left that winter in the care of Vincent Boucher, the foreman. Then, in one month during the summer of 1934, a few men were able to

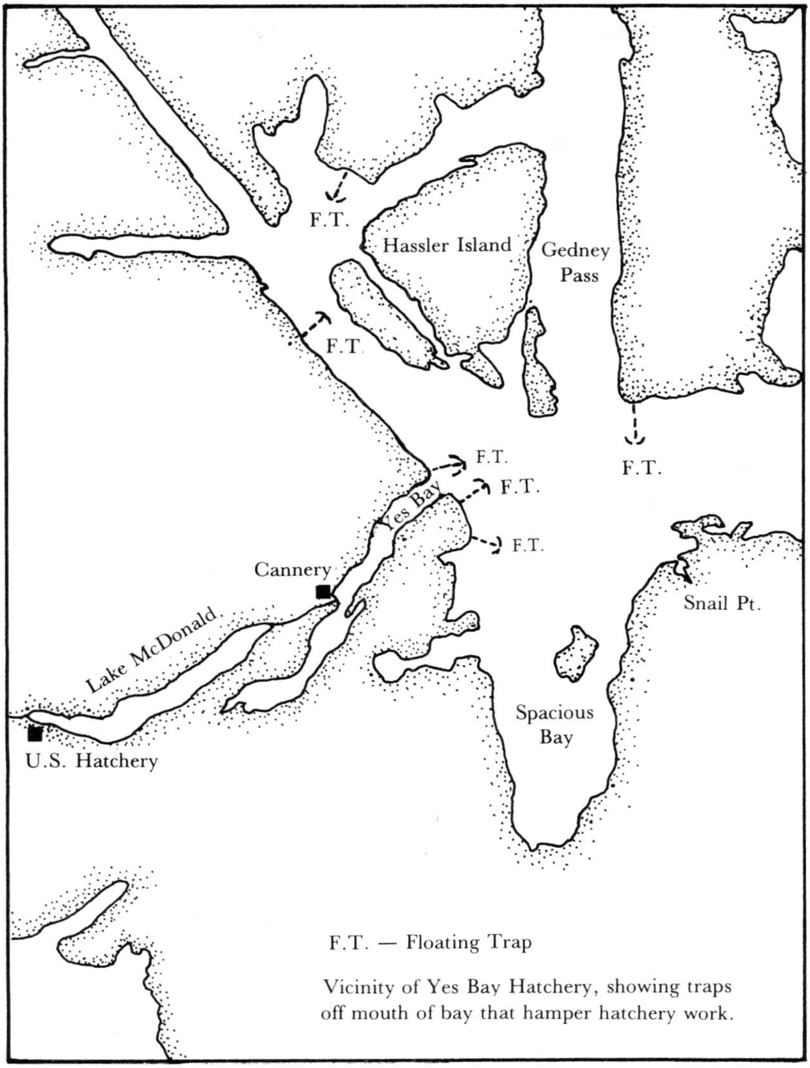

Map 11. Location of Fish Traps near Yes Bay, 1914 *(Bureau of Fisheries)*

dismantle the station, barge the equipment the length of the lake, haul it over the tramway and load it on ships. These same tasks in the reverse direction took many men a period of a quarter of a century to accomplish. The last entry in the station logbook dated September 24, 1934 was truly a death knell "This station is now closed and abandoned as far as the Bureau is concerned."[35]

Much of the property was shipped to the Bureau's fur seal stations on the Pribiloff Islands. Hatchery equipment such as the new yellow cedar troughs were transferred to Clackamas, Oregon and to other Bureau hatcheries. Two hundred and forty such troughs had been newly constructed in 1930-31. It took Ketchikan Spruce Mills nearly three years to gather 250 planks, 16 inches wide by 16 to 18 feet long, vertical grain, dressed and free of knots. The sides of the troughs were of the same specifications except 8 inches wide. The remainder of the property — the tramway, the pipeline, launch, scow, and boats, the donkey-house on the tramway, and all buildings — were turned over to the U.S. Forest Service which put in a Civilian Conservation Corps camp. The majority of the buildings were dismantled and the lumber taken to Ketchikan.

There was one last detail. Franklin D. Roosevelt revoked the executive order reserving the land and water for a fish propagation site on May 31, 1935. The area was declared part of the Tongass National Forest.

Estimates of the escapement of sockeye into the Yes Bay system have been inconsistent since the 1930s. Alaska Department of Fish and Game records show 10,000 sockeye were observed on August 30, 1964; 4,500 on August 21, 1967, and 15,000 on September 7, 1976.[36] A weir was maintained for the first time in 1981, and over 129,000 sockeye were counted.[37]

Afognak Hatchery

The second federally-funded hatchery to be constructed in Alaska began at Afognak Island in 1907. The area had been recognized earlier as a potential fish culture site, but it took twenty years for the site to be utilized for that purpose. The site was first examined in 1889. That year Tarleton H. Bean, the Commission of Fish and Fisheries ichthyologist, headed a group of scientists who went to Alaska under authority of an act approved March 2, 1889, to investigate the habits, abundance and distribution of Alaskan salmon and methods of fishing. Livingston Stone was detailed to join the group to inquire into the feasibility of fish culture. At that time Stone was superintendent of federal fish hatcheries in California and Oregon and had selected the site and built the Baird hatchery on the McCloud River in California in 1872.

Bean and his group visited Southwest Alaska and in particular Kodiak Island where the major fishing area was at the mouth of the Karluk River. Here Stone made the assessment that it would be difficult to sustain a hatchery at Karluk Lake because of the necessity of a road from Larsen's Cove on the opposite side of the island.[38] Other sites in the vicinity were visited and the one that impressed Stone was on Afognak Island near the falls of the Litnik River, now known as the Afognak River (Map 12). (The name Litnik has been retained throughout this chapter for the Afognak River and Afognak Lake.)[39]

Stone listed the reasons for his choice of this site for Alaskan fish culture operations:

> "1) a river that is easily controlled and not subject to excessive fluctuations of rise and fall,

2) an unlimited supply of water for the hatching house, furnished by gravitation,
3) favorable conditions for capturing the breeding fish,
4) a central and comparatively accessible location.''[40]

Afognak Island sparked Stone's enthusiasm and in 1892 he published an article, ''A National Salmon Park'' in *Forest and Stream*, and subsequently read it before the American Fisheries Society.[41] Stone's main thrust was that, like

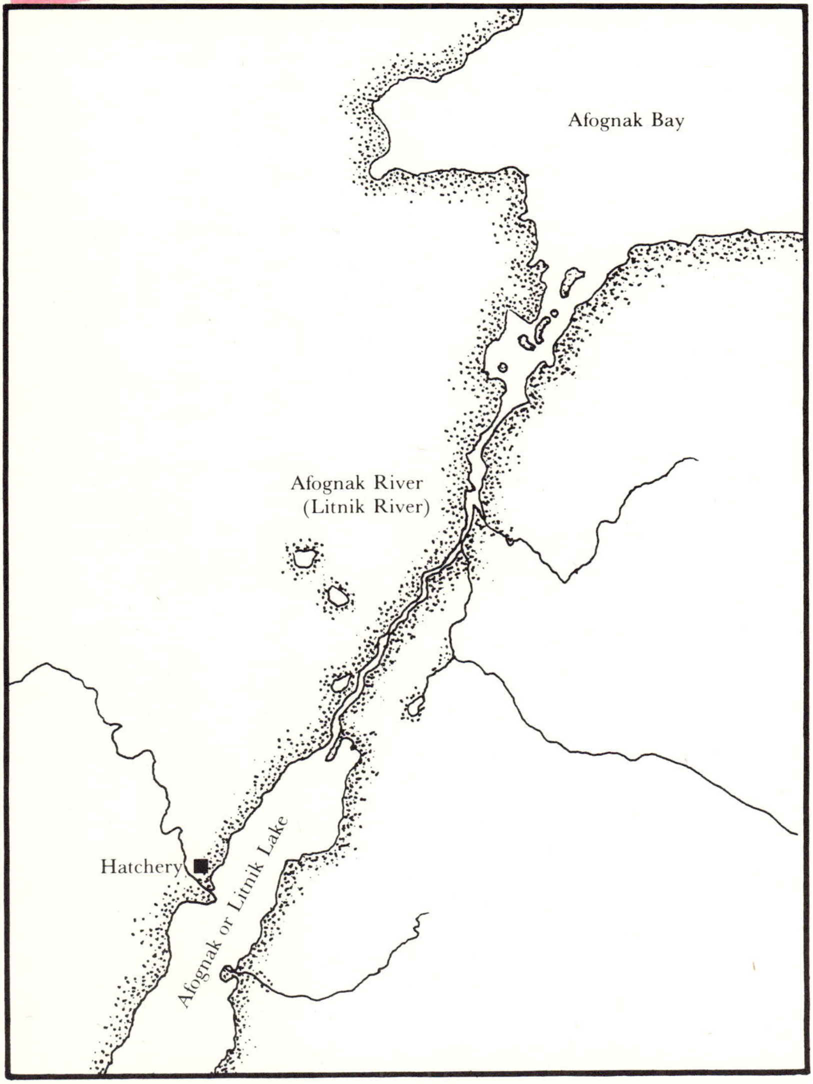

Map 12. Afognak (Litnik) Lake and River

the buffalo, the salmon were being destroyed. There was no refuge for the salmon, and he recommended that Afognak Island be set aside as a national park for salmon. Although Afognak Island was not declared a national park, it was set aside by Presidential Proclamation on December 24, 1892, as a Forest and Fish Culture Reserve. Also included were Sea Lion Rocks and Sea Otter Island. The reserve was for use of the United States Commission of Fish and Fisheries for fish cultural stations and was also the first forest reserve created in Alaska.

The proclamation ordered an immediate halt to commercial and subsistence fishing in the bays and waters surrounding Afognak Island to preserve the supply of fish until a hatchery was built. It soon became evident that the reservation caused many problems due to the failure to fix a penalty for trespass either upon the land or the streams of the island or waters surrounding it. No means of enforcement was provided.

The first to recommend abandonment of the reservation was Fisheries Agent J.K. Luttrell in 1893, only a year after the proclamation.[42] His opposition was based upon the argument that the whole island, 600 square miles, was not needed for fish propagation purposes. More outspoken was Fisheries Agent Howard M. Kutchin in the late 1890s. He frequently submitted proposals to the Commissioner of Fish and Fisheries to revoke the reservation order. Kutchin felt that distinct injury had occurred to the Natives who lived on the island and evils had developed because of flagrant fishing violations by large canning companies.[43] There was little Kutchin could do about the fishing violations except report them to his superiors and warn the companies involved. His concern for the Natives, who were not allowed to fish for their own use, prompted a revision in the regulations in 1900 which allowed subsistence fishing for the people of Afognak and Little Afognak villages. Natives also could salt fish for sale to purchase necessary supplies.

It was, therefore, a great disappointment, after years of disturbances and controversy, that the first congressional appropriation for fish culture in Alaska was spent on the Yes Bay hatchery in Southeast Alaska. Expediency was most important, and Afognak Island was bypassed. Once the Yes Bay hatchery was in operation, the second half of the original appropriation was committed to build a second Alaskan facility. Haste was no longer of utmost necessity and in 1907 the Afognak Reservation was dedicated to its original purpose.

Again, Claudius Wallich supervised construction and initial fish cultural activities. This time accompanied by a fish culturist from the Yes Bay hatchery, C.P. Henkel, Wallich began a similar search to the one he had undertaken at Yes Bay two years earlier. The pair traveled by commercial steamer to Kodiak and hitched a ride on a fishing boat to the mouth of the Litnik River, about thirty-five miles distant.

It did not take long for Wallich and Henkel to reject the site favored by Stone sixteen years earlier. A hatchery located at the base of the falls required that a method for holding adult salmon to maturity be developed and the Bureau was fully aware of the inadequacy of the ripening ponds at Karluk hatchery. The idea that green eggs and fry could be carried between the lake and a tidewater hatchery was immediately rejected.

Wallich and Henkel repaired two skiffs found in an old Alaska Packers Association building and dragged them to Litnik Lake, a distance of five or six miles. The lake is about five miles long by three-fourths of a mile wide and surrounded by mountains on all sides. Exploration showed two major streams entering the lake. The west stream had many disadvantages, including the distance which supply water would have to be flumed. Henkel selected a site on a stream on the east side of the lake, one mile from the outlet.[44] Here he found a plateau of perhaps five acres, a gravity feed water supply, and ideal spawning grounds with evidence of a fair run of sockeye salmon.

The construction crew and supplies arrived in June, 1907. The situation was similar to that at Yes Bay as transportation systems were the first necessity, but here the similarities ceased. Unlike the Southeastern location, no local supply of lumber was available and it was necessary to set up a sawmill and logging outfit. Within a week 27 men had steam in the boiler, the equipment set up, and 20,000 board feet of logs cut and rafted to the log haul. Five miles of tramway were completed from Afognak or Litnik Bay to the lake during the summer of 1907.[45] When this was completed, the sawmill was moved to the hatchery site and in mid-September 1907 actual construction of the buildings at the hatchery began. It was not until the following May that the water supply system and the hatchery equipment, such as troughs, were started.

The cost and time of construction exceeded expectations. Wallich told the Washington office, "Try as hard as I can, I have failed so far to uncover anything in connection with the Afognak station that is either reasonable or cheap."[46] He explained that everything except some of the lumber had to be transported five miles from the old cannery buildings at the bay to the hatchery. The tramway had wooden rails, despite the fact that Wallich had specified iron rails. It took 20 men to drag 390 feet of lumber over the inclines, and 34 men were required to move the boiler and equipment, even using block and tackle. Logging operations had been very expensive, as the yoke of oxen had been cut from the budget. Instead eight to ten men dragged the logs from the woods to the lake.

Despite the difficulties, the hatchery was completed in time for taking of spawn in the fall of 1908. This hatchery was a frame, one and a half story building, 197 feet by 55 feet, containing 240 Williamson troughs (Fig. 44). These troughs were 15 feet 9 inches by 14 inches by 6.5 inches (inside measurements), placed six each in double rows across the building (Fig. 45). Each trough had partitions for six baskets. Like the Yes Bay hatchery, the egg capacity was for 72 million eggs, but in only two of the 23 years that eggs were taken was capacity reached (Table 24).

Figure 44. Afognak hatchery complex, circa 1925. *Courtesy Harry D. Baer.*

Figure 45. Interior of Afognak hatchery, 1914. *Bureau of Fisheries.*

To provide a steady supply of water, a dam, 85 feet long and 16 feet high, was constructed on Hatchery Creek, above the rapids, giving a 13 foot fall. A 500 foot wooden flume, 12 inches by 24 inches, transported the water by gravity. The dam was duel purposed, as it served as a barrier to migrating fish. Thus, seining took place at the creek mouth. During later years fish were also captured by use of racks and seines from several small creeks at the head of the lake.

The first eggs taken in the fall of 1908 did not fulfill expectations as the run was very light. Only 46,380,000 sockeye, and 10,000 pink eggs were secured. Pink salmon were captured in the Litnik River and tributaries close to salt water. Once salmon were captured, standard spawning procedures were used. Water temperatures ranged from 32 °F to 37 °F average during winter months and from 40 °F to 55 °F average in summer.

Sufficient eggs were taken in 1909 to fill the hatchery, but the following year the sockeye run produced only a little over 30 million eggs. So in 1911 a field station was opened at Malina Lake, which empties on the side of Afognak Island, opposite the Litnik catchment basin. It was necessary to carry these eggs overland several miles, but it is not recorded if eggs were green or eyed.

Plans to develop the Litnik Lake run were disrupted by the Mount Katmai eruption on June 6, 1912. This eruption covered Afognak Island, 60 or more miles distant, with sand, ash and dust which varied in depth from ten inches on the south side to three inches on the northern portions. Streams, large and small, were literally choked with ashes and those which ran swiftly washed it to sea where, more often than not, the ash was deposited several feet in depth at their mouths.

Salmon in the streams during this time were driven back to salt water or perished in the mud. The mortality rate at Litnik Lake varies according to the source. Fisheries Agent E. M. Ball stated that the heaviest mortalities on Afognak Island were at the weir at Litnik Lake, where about 4,000 fish died.[47] The 1912 annual Alaska fisheries report, *Fisheries and Fur Industries in Alaska*, reported that 8,000 to 10,000 fish were killed below the rack during the day the ashes fell.[48] The Commissioner of Fisheries in his annual report estimated the number of dead salmon observed at the head of the lake at 8,000 and there may have been more covered by ashes.[49] This number may be in addition to the other figures given.

The eruption happened early in the season and few fish entered the streams by August. Nearly two months after the eruption, the streams and lake were still so muddy that any object two inches below the surface could not be seen. In addition the streams were very low, leading to a disinclination of salmon to ascend to the spawning grounds. Sufficient sockeye did enter Litnik Lake so that 14,689,470 eggs were taken.

Table 24. Sockeye Spawn Taken, 1908-1932: Afognak Hatchery

Brood Year	Number of Eggs Taken	Spawning Period	Comments
1908	46,380,000	July 27-Aug. 26	...
1909	76,020,000	July 30-Aug. 31	...
1910	30,725,000	July 30-Aug. 28	...
1911	30,520,000	July 21-Sept. 2	Includes several million eggs taken at Malina Lake.
1912	14,689,470	July 27-Aug. 31	Mount Katmai eruption — June 6.
1913	10,989,000	July 31-Oct. 10	Small return of fish.
1914	7,390,000	Aug. 1-Sept. 29	Severe storms caused thick ash in streams; fish returned to sea and reappeared in reduced numbers.
1915	6,353,000	Aug. 7-Sept. 30	Drought kept fish out of lake.
			Additional eggs transferred eyed to hatchery: 2,600,000 Uganik Lake; 3,173,000 Seal Bay; 15,000,000 Yes Bay.
1916	17,044,000	July 28-Oct. 18	Additional eggs transferred eyed to hatchery: 681,000 Uganik Lake; 4,600,000 Seal Bay; 1,016,000 Karluk River.
1917	53,036,000	July 30-Sept. 11	Transferred 18,000,000 eggs stateside.
1918	54,681,000	July 30-Oct. 2	Transferred 25,700,000 eggs stateside.
1919	79,178,000	July 26-Aug. 22	Transferred 10,000,000 eggs stateside. Estimated 50% of run spawned for hatchery use. 6,000,000 eggs taken Sept. 8 for experimental purposes.

The run did not materialize in 1913. The first sockeye appeared as usual in early May, but their movement into the streams was checked by the large quantities of volcanic ash still being washed from the hills by melting snow and heavy rains. Litnik Bay and River fared worse than any of the other localities and at one time it was feared no fish would appear.[50] When the rack across the stream was removed on July 7, 48 sockeye had entered the lake. Ordinarily this date would have marked the end of the season.

Fisheries Agent Ball expressed his views on the small 1913 run.[51] There was mostly sand in the stream in April 1913, which Ball conjectured would "hardly

Table 24 — Continued

Brood Year	Number of Eggs Taken	Spawning Period	Comments
1920	62,300,000	July 31-Sept. 24	Transferred 10,340,500 eggs stateside. High water prevented capture of adult salmon.
1921	53,835,000	July-Sept. 20	Transferred 5,200,000 eggs stateside. Floods prevented successful fishing; one-fourth of run used for hatchery purposes.
1922	61,790,000	Aug. 1-Sept. 14	Transferred 10,678,400 eggs stateside. Used half of run for hatchery purposes.
1923	No Eggs Taken	...	Run too small.
1924	No Eggs Taken	...	Run too small.
1925	11,000,000	Aug. 3-Sept. 15	...
1926	21,250,000	July 27-Sept. 10	Transferred 3,402,000 eggs stateside.
1927	4,225,408	Aug. 7-Sept. 28	...
1928	20,442,120	Aug. 1-Aug. 24	Transferred 1,851,550 eggs stateside. Small run reported to all Afognak Island streams.
1929	22,000,120	Aug. 5-Sept. 24	Transferred 4,553,200 eggs stateside.
1930	6,347,270	July 29-Sept. 10	...
1931	23,054,720	July 29-Oct. 3	...
1932	19,151,800	...	...

be considered more injurious to the spawning beds than the ordinary sediment and slime so universally present in these streams.''[52] He pointed out that there were an unusual number of coho present in many of the streams. His final analysis was that: ''. . . the scarcity of salmon in the streams and lakes can hardly be due to the condition of the spawning beds, but is likely caused by close fishing.''[53] Commercial fishing had been opened in the Afognak Reservation (except for Afognak Bay) for the first time in 1912. Nearly 101,000 sockeyes were caught in 1913 in the commercial fishery, and only 10,989,000 sockeye eggs were collected at the hatchery.

The eruption caused problems within the hatchery, owing to the accumulation of volcanic ash in the hatchery troughs. Great difficulty was experienced in holding the comparatively small number of eggs collected in 1912 and heavy losses of both eggs and fry resulted through suffocation. This happened again in 1913, by the continued influx of volcanic ash, a fresh deposit of which resulted from another, though less violent eruption in September of that year. This ash was so troublesome that part of the sac-fry had to be put out of the hatchery as soon as possible.

The Mount Katmai eruption and its effects undoubtedly influenced the decision to expand fish culture work in the Afognak-Kodiak region. It was decided to continue the Malina Lake transfer of eggs in 1912, but it was found that the spawning creek at the head of the upper lake was almost filled with volcanic ash. The outlet was so clogged that no fish could enter and the plan was dropped.

Two experimental field stations were built at Eagle Lake and Uganik Lake, both on Kodiak Island, the next year, 1913. Temporary hatching troughs were built at each place, and a barrier placed across the streams to obtain an accurate count of fish. The one at Uganik Lake could not be maintained because of a swift current. Sockeye eggs were taken, eyed and with the approach of cold weather, planted in suitable lagoons of Eagle and Uganik lakes.

No sockeye eggs were taken in 1914 at the field stations, and the Eagle Lake station was removed because of volcanic ash deposits. A little over two and one-half million pink eggs were taken at Uganik Bay, eyed at a new site at tidewater, and the eyed eggs shipped stateside to fill an order by the Bureau of Fisheries.

Investigations on other sockeye producing streams suggested that Seal Bay on the northeast coast of Afognak Island would provide sufficient eggs to warrant establishment of a field station. Both sockeye and pink eggs were eyed there and transferred to the Afognak hatchery, the same procedure used at Uganik Bay.

Transplanting eyed sockeye and pink eggs continued in 1915 and 1916. Then in 1917 only pink eggs were taken at Uganik Bay and the sockeye eggs from Seal Bay were planted eyed in Little Seal Bay Creek. The taking of spawn at remote sites was abandoned after that year.

A shipment of 15 million eyed sockeye eggs was made to Afognak from the Yes Bay hatchery in 1915 to further rehabilitate the Litnik Lake runs. A little over one million eyed sockeye eggs from the Karluk hatchery were transferred to the Afognak hatchery in 1916 when that facility closed. After that, no other stocks of sockeye were released into Litnik Lake.

The reason for discontinuing transfers of sockeye eggs to the system was that the sockeye runs in Litnik Lake began to increase dramatically. It was estimated in 1919 only fifty percent of the run was used for hatchery purposes, and 79 million eggs were taken. Only a fourth was used in 1921 and half in 1922. In none of these years did the egg collection afford an accurate index of the numbers of fish entering the lake, as only in exceptional seasons did conditions permit active fishing throughout the spawning season. Thus, many of the fish annually escaped to spawn under natural conditions.

Species other than sockeye were also spawned and incubated (Table 25). A pink salmon run in the Litnik River was utilized not only for enhancement of that run, but for export of eyed eggs to other Bureau of Fisheries hatcheries stateside. Some fry from pink eggs taken at Karluk River, Uganik Bay, and Seal Bay were introduced into the Litnik River. Twice coho salmon eggs were incubated and the fry released in the lake, but generally this species was excluded from the lake. During several years steelhead eggs were collected from the Karluk River by Afognak hatchery personnel. These were incubated and eyed eggs often transferred. Occasionally fry were reared and released at various spots on Afognak Island. All activities with these species, however, were incidental to the main emphasis on sockeye salmon.

Once spawn was taken, standard incubation procedures were used. Water temperature during December through March was uniformly 32 °F so hatching of the eggs was greatly prolonged, some taking three months. Prior to the installation in the 1920s of the stacked tray system, the majority of the sac-fry were planted soon after hatching, chiefly from February to June. Chamberlain criticized the system after a visit in 1911 saying: "They dump yolk sac-fry into the sluggish stream inhabited by trout and other predator species, on a bottom of vegetation, at a time when there are empty troughs in the hatchery."[54] The troughs held about 10 to 12 million sac-fry without the use of stacked trays which trebled the sac-fry capacity. Capacity for rearing to fingerling size was reported to be six million.[55]

Records do not indicate when the sac-fry were first held until the sac was absorbed, or when it became the policy to feed the fry. The first mention of fry being held was made by Fisheries Agent Ball in 1913.[56] He indicated that the reservoir, from which the hatchery water was obtained, was utilized as a nursery, but soon became overstocked. A battery of 12 rearing ponds was built in 1915, with water supplied from Hatchery Creek. When Harry Baer reported to work at the station in 1927, there was no sign of these ponds.[57] It would appear that in later years of operations, the fry capacity was dictated by the stacked tray capacity.

Only a small portion of the fry were held for feeding. Records are scanty on fry diet. Dolly Varden captured for predator control were salted as well as the carcasses of the spawned salmon. Preparation of these salted fish consisted of soaking in running water for 24 hours, removing skin and bones, and grinding.

Table 25. Pink, Coho, Steelhead, and Dolly Varden Production, 1908-1932: Afognak Hatchery

Brood Year	Species	Number of Eggs Taken	Number of Fish Released	Distribution
1908	pink	...	10,000	...
1909	pink	499,000	363,740	...
1910	pink	405,000	364,150	Released as sac-fry.
1911	pink	6,472,000	5,829,300	Released as sac-fry.
1911	coho	244,700	214,000	...
1912	pink	3,271,740	151,000	Unknown if eggs transferred.
1913	pink	13,900,000	12,034,399	Released in May.
1913	coho	74,600	50,000	Released in Feb.
1914	pink	16,508,600	139,460	Eggs sources: 6,574,600 Litnik River; 2,534,000 Uganik Bay; 7,400,000 Karluk River. A shipment of 12,500,000 eggs transferred stateside.
1915	pink	14,958,000	13,104,200	Released as sac-fry. Eggs sources: 11,435,000 Litnik River; 2,400,000 Uganik Bay; 1,123,000 Seal Bay.
1916	pink	38,451,000	19,343,000	Egg sources: 25,310,000 Litnik River; 10,321,000 Uganik Bay; 2,820,000 Seal Bay. A shipment of 16,202,752 eggs transferred stateside.
1917	pink	2,860,000	1,326,000	Egg sources: 300,000 Litnik River; 2,560,000 Seal Bay, a portion of which was sent eyed to hatchery.

Table 25 — Continued

Brood Year	Species	Number of Eggs Taken	Number of Fish Released	Distribution
1918	pink	8,697,000	2,142,000	Transferred 5,760,000 eggs stateside.
1922	pink	600,000	240,000	Transferred 278,616 eggs stateside. Release of fry in tributary of Litnik River.
1926	pink	4,212,000	...	Transferred 3,617,000 eggs stateside.
1926	steelhead	2,060,000	850,000	Transferred 1,023,360 eggs stateside. Remainder released in local waters.
1927	steelhead	1,090,000	0	Transferred stateside.
1928	pink	2,300,800	0	Transferred 1,851,550 eggs stateside.
1930	pink	13,671,670	1,145,334	Transferred 10,155,776 eggs stateside. Remainder released in Litnik River in April and June.
1930	steelhead	123,904	110,000	As #1 fingerlings in Island Lake in July.
1931	steelhead	32,826	0	Planted as eyed eggs in Eagle Creek in June.
1931	Dolly Varden	2,731,424	0	Egg source: Buskin River, Kodiak. Transferred stateside.
1932	pink	2,524,500	0	Transferred 2,228,000 eggs stateside.

This mash was placed in boiling water which caused the flesh to granulate and the mash was then reground. When granulated by this treatment, a greater proportion was eaten, whereas, when the raw flesh was reduced to a fine state by the second grinding, a large part, in the form of milky fluid, was lost when put in water. The fry were fed from six to eight weeks and released when it became necessary to provide feeding space for more recently hatched fish. Unlike other stations where salt fish were used, Afognak hatchery reported that the results of such feeding were satisfactory.[58]

The majority of the sockeye fry produced at the hatchery were distributed in Litnik Lake, well out in the middle, where it was believed there were probably fewer predators (Table 26). From the late 1920s until the closure of the station, it became the policy to release about two million fry in each of the two little lakes at the head of and on the south side of Litnik Lake. None were placed in the three lakes tributary to the lower end of the lake. Willis H. Rich, a professor at Stanford University who undertook studies for the Bureau of Fisheries, examined the lakes in 1930. He felt that it was impossible for the fry to find their way from these lakes through such insignificant 12 to 18 inch wide outlets, unless it was possible during spring freshets. He stated that the extent of the lakes ". . . is so small that the food supply cannot be great and conditions are so unnatural for young salmon that it might well be that the mortality is much above that in the main lake."[59] He went on to advise the superintendent of the hatchery, "Until proven otherwise, it would seem much more desirable to plant the fry in the main lake except for experimental purposes and if one were to experiment with plantings in these little lakes, the experiments should be closely watched."[60] No records survive which indicate whether these plantings were discontinued.

It is not known if personnel released sockeye salmon except in the main lake and these small lakes. Rich learned in 1930 that some sockeye fry were released in a small lily covered pond, known as Aleut Town Lake, near the Afognak village in 1925 or 1926.[61] One of the employees caught some adults off the inlet in 1930.[62] Harry Baer, a former employee, remembers that: "Only a bucketful of salmon fry were planted in the little lake, and there was quite a little school of adult salmon trying to get back up the little ditch that comes from the lake, but couldn't make it and wouldn't have had suitable spawning grounds if they had made it."[63]

There was concern about the condition of Litnik Lake and its ability to provide food for the fry released into it so limnological studies were contracted to Rich in 1930. He found a dearth of plankton in the lake and in 1931 samples of young fish showed they were decidedly poorly nourished. This was brought to the attention of the Commissioner of Fisheries, Henry O'Malley, who corresponded with Dr. Chancey Juday, University of Wisconsin, about fertilizing Litnik Lake with phosphate. O'Malley said, ". . . it seems to me as years go by [lake fertilization] is one of the most valuable things we can do. There is certainly no sense in planting fish in waters that are devoid of food, for the fish simply

Table 26. Sockeye Released, 1908-1932: Afognak Hatchery

Brood Year	Fish Released	Distribution
1908	39,325,870	Unknown spots in Litnik Lake.
1909	68,422,170	Unknown spots in Litnik Lake.
1910	26,755,000	Unknown spots in Litnik Lake.
1911	18,394,700	Unknown spots in Litnik Lake.
1912	12,551,100	Released principally in February and March in places about Litnik Lake where streams were not awash with ash from Mount Katmai eruption.
1913	7,761,700	Unknown, from February through June.
1914	5,044,830	Unknown, as fry from December-May; 942,250 as sac-fry.
1915	22,933,640	Unknown spots in Litnik Lake; combined release from eggs taken at Uganik, Seal Bay, and Yes Bay. Sac-fry and fingerlings planted February-June. 11,100,490 fed to fingerling stage.
1916	21,116,000	Combined release into Litnik Lake from eggs taken at Uganik, Seal Bay and Karluk. 1,401,000 sac-fry planted October-November, 1916. 10,160 released as fingerlings.
1917	31,427,000	Litnik Lake and tributaries.
1918	25,583,000	Unknown spots in Litnik Lake.
1919	61,524,000	Unknown spots in Litnik Lake.
1920	47,808,000	Unknown spots in Litnik Lake.
1921	32,580,000	Disease epidemic in May and June, large loss of fry. 28,480 fingerlings and 4,100,000 sac-fry released in local waters.
1922	46,333,000	All fry held until sac absorbed. 30,000,000 reared and released in July.
1923	0	No releases made.
1924	0	No releases made.
1925	10,075,000	As #1 fingerlings in Litnik Lake and tributaries.
1926	14,400,000	As #1 fingerlings in Litnik Lake and tributaries.
1927	3,675,000	As fingerlings in Litnik Lake and tributaries.
1928	18,700,000	As #1 fingerlings in Litnik Lake and tributaries.
1929	15,100,000	As #1 fingerlings: 11,880,000 and 3,120,000 as fry. Released in June in Litnik Lake and tributaries.
1930	5,553,080	As #1 fingerlings in Litnik Lake and small lakes at head of Litnik Lake.
1931	19,920,712	As fry from March-June, possibly in nearby lakes.
1932	17,400,000	As fry, unknown distribution.

carcasses were removed then they wanted to fertilize.

starve to death in a short time.''[64] Juday put together a plan, but the project was delayed for lack of funds and then abandoned when the station closed.

Scientific evaluation of the success of the hatchery program at Afognak was negligible. On three different occasions attempts to monitor outmigrants were made but none proved successful. The first attempt was in 1913, the spring after the Mount Katmai eruption. Collecting stations were established at the foot of the lake and in the bay, about two miles below where the stream entered salt water. The meager number of young salmon justified the conclusion that there had been a heavy loss.[65] No description of the methods of capture or statistical analysis have survived.

The use of traps to check outmigrant fry was tried in May 1928, but high water prevented effective operation. Visual attempts to estimate numbers of migrating sockeye fry were tried in May 1929, but again high water and the roily condition of the water made such estimates impossible.

A rack or weir was built in 1920 on the Litnik River. However, its purpose was not to check the efficiency of the hatchery, but rather to count returning sockeye salmon to determine if commercial fishing should be opened in Afognak Bay. By use of the rack it was possible to ascertain if adequate numbers of fish had entered the lake to ensure a sufficient number of eggs for propagation. A partial count of fish through the rack was kept (Table 27).

Harry Baer, who worked at the hatchery from 1927 to 1932, was puzzled when shown the weir records on pink salmon. ''The counting weir was only a short distance below the exit from the lake, and I never did see a pink salmon up as far as this counting weir, or in the lake, so I can't understand the counts at the weir of pink salmon, unless the counter was confusing sockeye for pinks.''[66]

The counts for coho salmon were not always accurate, as some years the weir was closed in August to exclude that species from the lake. Cohos were considered predatory on sockeye fry, but the major predator was the Dolly Varden.

A constant battle to eradicate Dolly Varden continued throughout the life of the hatchery. The first attempt to destroy this predator appears to have started in 1911 when a rack barricaded the Dolly Varden from the lake, but only 20 were taken and the rack washed out. So in the fall of that year a seine was utilized at the outlet to the hatchery flume, where the sockeye fry were often released. One haul netted so many Dolly Varden they filled a small dory. Fish traps set during the winter produced several hundred Dolly Varden. The Dolly Varden population was reported scarce by 1913.[67]

This condition did not last and during the spring of 1919, the hatchery stream was teeming with Dolly Varden of uniform size and weight (average length, 10 to 12 inches and weight, 8 to 10 ounces). Their stomachs were reportedly gorged with sockeye fry.[68] Drag seines and traps continued to be

Table 27. Litnik (Afognak) River Weir Counts and Afognak Reservation Harvest Records, 1912-1933

Year	Litnik River Weir Counts			Afognak Reservation Harvest Counts		
	Sockeye	Pink	Coho	Sockeye	Pink	Coho
1912	...	...	...	82,601	46,960	8,525
1913	3,367	6,882		100,988	47,673	4,113
1914	...	...	...	144,568	63,148	3,309
1915	...	...	...	91,289	37,494	5,938
1916	...	...	...	46,311	5,470	21,267
1917	...	...	...	71,527	107,333	3,558
1918	...	...	...	50,662	70,791	5,151
1919	...	...	...	79,097	32,998	10,417
1920	...	...	...	52,397	60,944	12,162
1921	37,653	...	...	173,443	11,892	6,136
1922	...	...	...	42,149	94,072	36,437
1923	8,025	...	...	146,761 (a)	96,853	24,334
1924	10,317	15,824	12,761	102,056	41,799	36,714
1925	11,000 (b)	...	...	106,739	76,038	30,969
1926	22,250	4,646	1,120 (c)	33,003	110,344	153,434
1927	7,491	1,075	4,216 (c)	28,844	174,685	32,821
1928	20,862 (d)	144 (c)	833 (c)	34,011	237,519	45,483
1929	25,428	227 (c)	435 (c)	STATISTICS HEREAFTER PART OF KODIAK DISTRICT		
1930 (e)	6,238	6,015	136			
1931	30,515	234	3,534			
1932	23,574	5,642	654			
1933	36,144 (f)	...	...			

(a) Litnik Bay opened for the first time to commercial fishing and 758 sockeye taken.
(b) Estimates.
(c) Weir closed in August to exclude coho.
(d) Natives took 500 sockeye for subsistence before weir set up.
(e) Weir not fish-tight.
(f) Natives took 3,000 sockeye for subsistence before weir set up.

used. Between 6,000 and 7,000 were captured in 1924; 54,100 in 1925; 35,950 in 1926 and 20,000 in 1928. Persistent seining for four years virtually cleared the lake of the predator, but again this condition did not last and seining continued until the station closed in 1933.

The sockeye productivity of Litnik Lake prior to the establishment of the hatchery is surprisingly obscure considering its reserved status since 1892. Prior to the purchase of Alaska, the Russians built a dam or *zapor* in the river which was maintained by their descendents, first for their own use and later to supply two canneries in the area.[69] These canneries, built in the late 1880s at the head of Afognak Bay, were three-quarters of a mile below the river mouth. Both had crews fishing in 1889 and 1890 and one fished in 1891, but the pack was not totally from Litnik River fish, so examination of their pack is not enlightening.

Early federal fisheries people attempted to estimate the run. Stone describes the run as "fairly abundant" in 1889; in comparison to the Karluk run which was "extremely abundant."[70] The U.S. Deputy Surveyor for Alaska, A. Lasey, while surveying the area for the Afognak Reservation, learned that

> "The numbers of red salmon that endeavor to ascend the Afognak River to spawn in the lake is, owing to rapids and natural obstructions, comparatively small. The Karluk river and lake being the great and natural breeding grounds of the red salmon. The greater part of the fish caught in Afognak Bay are passing schools."[71]

Moser felt the stream had produced enough sockeye for 10,000 to 15,000 cases a year prior to 1897.[72] Evermann, using Moser's figures and estimating 13 fish to a case, gave the total output of 130,000 to 195,000 fish.[73] But Moser revised his estimate in 1901, stating that the stream would furnish without damage to the run 100,000 to 150,000 sockeyes each year.[74] It is unknown where Moser acquired his data. Presumably it was from estimates made by cannery operators who had fished prior to the 1892 proclamation, which closed the waters around Afognak Island. The effect of this closure was to encourage a spirit of lawlessness which rendered the Executive Order not only inoperative, but made its provisions an object of contempt and the efforts to enforce them an instrument of derision, as one agent put it.[75] It was impossible to prevent taking of salmon and, with no penal clause, the sole remedy for persistent violation was to capture trespassers in the act, eject them, and seize their fishing gear. As one agent said, "How impractical this was none knew better than the trespassers."[76] The Bureau of Fisheries did not have sufficient men stationed in Alaska for a patrol nor a vessel at their disposal.

Fishing continued surreptitiously. The Natives took small numbers of fish for their own needs legally after 1900. No record was kept of any of these catches. Thus, when the hatchery was built in 1907-1908 on the Afognak Reservation, no one knew the status of the runs. It was assumed by those in the Bureau's Washington office, that the run which had been protected for 15 years was a strong one — one which would provide 72 million sockeye eggs to fill the hatchery to capacity. The first year only 46,380,000 eggs were obtained.

Despite the small numbers of eggs available three of the first four years of operations, political pressure was such that commercial fishing in the Afognak Reservation became legal in 1912 to the Natives of Afognak, Spruce, Whale, Dry and Raspberry islands. Fishing was by license under supervision and control of the Department of Commerce and Labor, who controlled Alaskan fisheries. Later this fishing privilege was extended to white persons who were permanent residents of the islands' villages. Harvest statistics were kept until 1929 when the reservation fishing was placed under control of the Kodiak area management (Table 27).

The status of the run into Litnik Lake in 1912 when commercial fishing began was fairly low. That same year came the June 6 Mount Katmai eruption. The effect on the hatchery and runs has been discussed earlier. When the runs did not improve in the years after the eruption, sockeye eggs were transferred in 1915 and 1916 to the Afognak hatchery for release as fry into the system (Table 26). Thereafter, from 1917 to 1922 runs were so large that returning adults were utilized.

Then in 1923 an exceptionally small number of sockeye ascended the spawning grounds adjacent to the station. It was decided to defer active salmon propagation for one year and devote funds to repairs. Sockeye harvest records within the reservation during that year show 146,751 fish or the second largest take since commercial fishing began in 1912.

In his fiscal report the superintendent wrote that it was impossible to assign a definite cause for the material decline from the larger runs.[77] However, it was ascertained that many sockeye were captured in the course of the spawning season near the mouth of the river at Little Afognak Bay. These fish proved to be similar in size and it was concluded that at least part of the shortage in the run could be reasonably ascribed to this fact. This phenomenon was first recorded in 1916 by Fisheries Agent E.M. Ball who wrote in his annual report:

> "At Little Afognak, the catch shows an increase of more than 200 percent, and that two-thirds of the salmon taken from Afognak waters came from that field alone. It shows further a marked change in the character of the run. Previously a large proportion of the red salmon taken at Little Afognak were undersized, and were accepted at the cannery as half-salmon only, two being counted for one. This year the type of salmon was markedly changed, the percent of small fish falling far below that of other seasons. Opinions are unanimous that this changed and increased run of salmon is the first unmistakable evidence of the effect of artificial propagation in this region."[78]

Again in 1924 the small number of returning sockeye to Litnik Lake did not warrant hatchery operations. Harvest records for that year show 102,056 sockeye salmon, the fifth largest in 12 years. For the third year in a row, in 1925, the run was small and the harvest was large, 106,739 sockeye.

A small run was expected in 1926 because disease had been epidemic on the 1921 brood fish and later the yearlings in the lake became affected and perished in large numbers. A description of this epidemic may be found in Chapter II. Despite the anticipated small return, the sockeye salmon counted through the weir was the largest number since 1924, but commercial catches were down drastically to 33,003 sockeye salmon. Fish traps, anticipated to intercept salmon which ordinarily would have gone to the Afognak Island streams, were first constructed during the season of 1927 on the north end shore of Raspberry Island.

After that time the large runs of sockeye salmon did not materialize again during the life of the hatchery. The Afognak station closed at nearly the same time as the Yes Bay hatchery, after the visit of a new Commissioner of Fisheries, Frank T. Bell. The reason for the closure was given as "economy requirements and some question as to the value of operations."[79] The Bureau of Fisheries was ready by August 13, 1934 to take final steps to dispose of or transfer the Afognak station to other government agencies. Ultimately it was transferred to the U.S. Navy who used it as a recreational center until 1968-1969.

In more recent years the sockeye runs to Litnik or Afognak Lake have been given by the Alaska Department of Fish and Game at 52,677 in 1978 and 84,702 in 1979.[80] Harvest records are not included.

Notes

1. Memo from H.C. Fassett to the Commissioner of Fisheries dated March 29, 1905, File "Yes Bay, 1905-1906," Records Concerning Fish Culture Stations, 1875-1929, Division of Fish Culture, RG 22, NA.

2. Ibid.

3. Memo from Barton Evermann to the Commissioner of Fisheries dated March 29, 1905, File "Yes Bay, 1905-1906," Records Concerning Fish Culture Stations, 1875-1929, Division of Fish Culture, RG 22, NA.

4. U.S. President (T. Roosevelt), "Executive Order, February 1, 1906," Yes Bay Station, Division of National Fish Hatcheries, U.S. Fish and Wildlife Service, United States Department of Interior, Washington D.C.

5. Within six months of selection of the hatchery site, Washington officials no longer used the name Yes Lake, but called it McDonald Lake so named for Marshall McDonald, U.S. Commissioner of Fish and Fisheries 1888-1895.

6. Claudius Wallich, "A Method of Recording Egg Development for Use of Fish Culturists," *Report of the Commissioner of Fish and Fisheries for the Year Ending June 30, 1900* (Washington: Government Printing Office, 1901), pp. 185-195.

7. Letter from Claudius Wallich to the Commissioner of Fisheries dated June 1, 1905, File "Yes Bay, 1905-1906," Records Concerning Fish Culture Stations, 1875-1929, Division of Fish Culture, RG 22, NA.

8. Ibid.

9. Letter from W.A. Hancock to the Commissioner of Fisheries dated November 6, 1905, File "Yes Bay, 1905-1906," Records Concerning Fish Culture Stations, 1875-1929, Division of Fish Culture, RG 22, NA.

10. Upper troughs averaged 28 quarts of water each per minute and the lower troughs 20 quarts, which equaled 960 gallons per minute. The overflow was calculated at 40 gallons, thus, 1,000 gallons were used per minute. *(Yes Bay Logbook, July 1, 1912-July 1, 1917* (Alaska Historical Library, Juneau, Alaska), March 14.

11. Letters from W.A. Hancock to the Commissioner of Fisheries, years 1908-1909, File "Correspondence of the Commissioner of Fish and Fisheries," Records Concerning Fish Culture Stations, 1875-1929, Division of Fish Culture RG 22, NA.

12. Letter from H. von Boyer to W.A. Hancock dated May 11, 1907, File "Correspondence of the Commissioner of Fish and Fisheries," Records Concerning Fish Culture Stations, 1875-1929, Division of Fish Culture, RG 22, NA.

13. Ibid.

14. Statement of Alphonse Kemmerich, former superintendent, in personal interview, Woodburn, Oregon, July 30, 1979.

15. Ward T. Bower and Harry Clifford Fassett, *Alaska Fisheries and Fur Industries in 1913,* Bureau of Fisheries Document No. 797 (Washington: Government Printing Office, 1914), p. 69.

16. Ibid, p. 68.

17. *Yes Bay Logbook, July 1, 1912-July 1, 1917* (Alaska Historical Library, Juneau, Alaska), May 23, 1914.

18. Ward T. Bower and Henry D. Aller, *Alaska Fisheries and Fur Industries in 1916,* Bureau of Fisheries Document No. 838 (Washington: Government Printing Office, 1917), p. 32.

19. *Yes Bay Logbook, January 1920-May 31, 1924* (Alaska Historical Library, Juneau, Alaska), May 9, 1923.
Kemmerich, July 30, 1979.

20. Ward T. Bower and Henry D. Aller, *Alaska Fisheries and Fur Industries in 1914,* Bureau of Fisheries Document No. 819 (Washington: Government Printing Office, 1915), p. 21.

21. *Yes Bay Logbook, 1906-1907* (Alaska Historical Library, Juneau, Alaska), May 22, 23, 1909.

22. Bower and Fassett, *Alaska Fisheries and Fur Industries in 1913,* p. 70.

23. Kemmerich, July 30, 1979.

24. Bower and Fassett, *Alaska Fisheries and Fur Industries in 1913,* p. 70.

25. Ibid.

26. Kemmerich, July 30, 1979.

27. *Yes Bay Logbook, 1907-1912* (Alaska Historical Library, Juneau, Alaska), p. 268.

28. Jefferson Moser, *Salmon Investigation in 1900 and 1901* (Washington: Government Printing Office, 1902), p. 294.

29. Ibid.

30. Howard M. Kutchin, *Report on the Salmon Fisheries of Alaska, 1900* Senate, 56th Congress, 2nd Session, Document No. 168 (Washington: Government Printing Office, 1901), p. 19.

31. Ibid.

32. Willis H. Rich and Edward M. Ball, "Statistical Review of the Alaska Salmon Fisheries, Part IV: Southeastern Alaska," *Bulletin of the United States Bureau of Fisheries Vol. XLVII* (Washington: Government Printing Office, 1935).

33. Letter from Ward T. Bower to Commissioner of Fisheries dated July 29, 1912, File "Correspondence of Ward T. Bower, 1911-1918," Reports and Related Records, Division of Alaska Fisheries, RG 22, NA.

34. Ibid.

35. *Yes Bay Logbook, October 1, 1932-September 19, 1934* (Alaska Historical Library, Juneau, Alaska), September 19, 1934.

36. John Edginton and Robert Larson, *Revised Anadromous Stream Catalog of Southeastern Alaska, Appendix A-District 1, Volume III, Western Behm Canal and Ketchikan Area Survey Data,* ADF&G Technical Data Report No. 30 (Juneau: Alaska Department of Fish and Game, 1977), p. 347.

37. Statement of Bob Campbell, Lake Fertilization Coordinator, Southern Southeast Regional Aquaculture Association, August 25, 1981.

38. Tarleton H. Bean, "Report on the Salmon and Salmon Rivers of Alaska, with Notes on the Conditions, Methods, and Needs of the Salmon Fisheries," *Bulletin of the U.S. Fish Commission, Vol. IX for 1889* (Washington: Government Printing Office, 1891), p. 207.

39. The name Litnik comes from the Russian word "Elitnik" meaning a place where fish are dried and prepared for future use. The spelling "Letnik" has also been used.

40. Bean, p. 208.

41. *Forest and Stream,* June 16, 1892.

Marshall McDonald, "Report on the Salmon Fisheries of Alaska,' *Bulletin of the United States Fish Commission, Vol XII for 1892* (Washington: Government Printing Office, 1894), pp. 14-19.

42. J.K. Luttrell, "Report of Special Agent Luttrell for 1893," bound in *Seal and Salmon Fisheries, General resources of Alaska, Vol II, 1898* (Washington: Government Printing Office, 1894).

43. Howard M. Kutchin, *Report of the Salmon Fisheries of Alaska-1898* (Washington: Government Printing Office, 1899), pp. 14-15.

44. This creek, known as Ahuyon Creek in offical reports, was called Falls Creek according to Harry Baer when he worked there from 1927-1932. On present maps it is called Hatchery Creek.

45. A road four and one-half miles long was completed November 30, 1929. Building of the road was financed and supervised by the Bureau of Public Roads. Its completion allowed use of a truck.

46. Letter from Claudus Wallich to the Commissioner of Fisheries dated September 10, 1907, File "Afognak, 1906-1909," Records Concerning Fish Culture Stations, 1875-1929, Division of Fish Culture, RG 22, NA.

47. E.M. Ball, "The Salmon Fisheries of Afgonak Island, Alaska in 1912," Reports and Related Records, Division of Alaska Fisheries, RG 22, NA.

48. Fred M. Chamberlain and Ward T. Bower, "Fishery Industries," *Fishery and Fur Industries of Alaska in 1912,* Bureau of Fisheries Document No. 780 (Washington: Government Printing Office, 1913), p. 34.

49. U.S. Bureau of Fisheries, *Report of the Commissioner of Fisheries for the Fiscal Year 1912 and Special Papers,* Document No. 772 (Washington: Government Printing Office, 1914), p. 43.

50. Bower and Fassett, *Alaska Fisheries and Fur Industries in 1913.* p. 58.

51. E.M. Ball, "Report on the Operations of the Inspector of Alaska Salmon Fisheries in 1913," Reports and Related Records, Division of Alaska Fisheries, RG 22, NA.

52. Ibid.

53. Ibid.

54. F.M. Chamberlain, "Report of the Work in Bristol Bay and Elsewhere in Alaska, 1913," Reports and Related Records, Division of Alaska Fisheries, RG 22, NA.

55. Willis H. Rich, "Notes from the Diaries of Willis H. Rich, 1922-1933" (National Marine Fisheries Service, Auke Bay Fisheries Laboratory library: Auke Bay, Alaska).

56. Ball, "Report on the Operations of the Inspector of Alaska Salmon Fisheries in 1913."

57. Letter from Harry D. Baer to Patricia Roppel, September 8, 1979.

58. Dennis Winn, "Season's Report, 1925, Southeast Alaska," Reports of Bureau of Fisheries Agents, Division of Alaska Fisheries, RG 22, NA.

59. Rich, "Notes From the Diaries of Willis H. Rich, 1922-1933."

60. Ibid.

61. Ibid.

62. Ibid.

63. Baer, September 8, 1979.

64. Letter from Commissioner of Fisheries Henry O'Malley to Dr. Chancy Juday, dated March 11, 1933, File "Correspondence with the Commissioner of Fisheries, Afognak Island, 1924-25," General Records, Division of Alaska Fisheries, RG 22, NA.

65. Bower and Fassett, *Alaska Fisheries and Fur Industries in 1913*, p. 65.

66. Baer, September 8, 1979.

67. Ball, "Report on the Operations of the Inspector of Alaska Salmon Fisheries in 1913,"

68. Ibid.

69. A zapor was a dam of saplings with stone piers across a stream. It was used by the Russians in Alaska to secure salmon prior to U.S. purchase. For a complete description see Moser, *Alaska Salmon Investigations in 1900 and 1901*, p. 244 and plate XX.

70. Bean, pp. 207-208.

71. Letter from A. Lasey to J. K. Luttrell, Department of the Treasury, "Letters Received Relating to Salmon Fisheries: 4/20/89 to 12/20/97," Alaska Files of the Secretary of the Treasury, 1868-1903. Microfilm 720, Roll 19, National Archives.

72. Moser, *The Salmon and Salmon Fisheries of Alaska*, p. 163.

73. Barton Warren Evermann and F.M. Chamberlain, "General Administrative Report," *Alaska Fisheries and Fur Industries in 1911*, Bureau of Fisheries Document No. 766 (Washington: Government Printing Office, 1912), p. 9.

74. Moser, *Alaska Salmon Investigations in 1900 and 1901*, p. 243.

75. Kutchin, *Report on the Salmon Fisheries of Alaska, 1900*, p. 42.

76. Ibid.

77. Glen C. Leach, *Propagation and Distribution of Food Fishes, Fiscal Year 1924*, Bureau of Fisheries Document No. 978 (Washington: Government Printing Office, 1925), p. 383.

78. E.M. Ball, "Report on the Fisheries of the Central District of Alaska, 1916," Reports and Related Records, Division of Alaska Fisheries, RG 22, NA, p. 31.

79. Glen C. Leach and M. James, *Propagation and Distribution of Food Fishes, Fiscal Year 1934* (Washington: Govenment Printing Office, 1935), pp. 385-386.

80. Letter from Roger C. Blackett, Alaska Department of Fish and Game fishery biologist, Kodiak to Patricia Roppel, October 8, 1979.

VI. The Territorial Fish Commission

Juneau

The Territory of Alaska's first salmon hatchery was the outgrowth of a trout enhancement program instigated in Juneau by A.J. Sprague, who became one of the most outspoken and controversial men in artificial propagation in Alaska. Sprague came to Alaska in 1907 from Leadsville, Colorado as an employee of the Bureau of Fisheries assigned to the Yes Bay station. At that time sockeye sac-fry were released almost immediately after hatching. This led Sprague to comment years later, "I first got the inside hunch of the failure and breakdown of all our salmon hatcheries."[1] Disillusioned because he saw no future for his belief that sac-fry should not be immediately released, Sprague left the Bureau of Fisheries and Alaska after one year.

He secured the position of superintendent of the Trask River hatchery near Tillamook, Oregon in 1909. He immediately started a pond rearing system, of which he wrote: "I first found that the ponds were so costly to build, they would not justify the great expense involved and began to look for a more natural means."[2] Sprague became convinced this meant planting eyed salmon eggs. He was never one to mince words and he wrote an article for the local Tillamook newspaper attacking the methods in use at Oregon hatcheries. This did not, however, produce the results he expected: "That finished me with the Oregon State Fish Commission. It was a little bit too early for them to learn the truth."[3]

He was able to return to the Oregon State Fish Commission within two years, this time at Klamath Falls. Here in 1912-1913 he organized the Klamath Falls Sportsmen's Club and erected a hatchery on Spencer Creek, a tributary of the Klamath River. He introduced eastern brook trout there, but became involved in a political dispute over jurisdiction of salmon and trout streams and again lost his job.

Back in Alaska, sometime between 1914 and 1916, Sprague met B.L. Thane of the Alaska Gastineau Mining Company at Juneau. Thane, Charles Garfield, and Sprague became interested in stocking barren lakes with trout. The trio examined the artificial reservoir constructed by the mining company at Salmon Creek for the hydroelectric power plant and Upper and Lower Annex lakes in Taku Inlet.

Application was made in October, 1916 to the Commissioner of Fisheries for an allottment of eastern brook trout eggs, Sprague's favorite species. With reasonable assurance the order would be expedited, Sprague began construction of a small hatchery on Annex Creek, Taku Inlet, under patronage of the Alaska Gastineau Mining Company. Construction commenced but an early cold snap, accompanied by much snow and inclement weather, indicated the site was not suitable. The equipment was transferred to the town of Thane where

Figure 46. A.J. Sprague, first Territorial fish commissioner for Alaska. *Nelson-Sprague collection, Alaska Historical Library.*

a small scale hatchery was constructed. When temperatures were 14 degrees below zero on January 13, 1917, 160,000 eastern brook trout eggs from the Bureau of Fisheries' hatchery at Leadsville, Colorado arrived in good condition. Over 150,000 eggs hatched and the resultant fry were released in June and July.

Sprague, Garfield, and B.L. Thane felt some organization should be formed to protect the planted trout and observe progress, so on May 24, 1917 a few Juneau sportsmen and public-spirited citizens organized the Alaska Fish and Game Club with Charles Goldstein as president. Over 300 of the business and professional men of the Territory joined the group. B.L. Thane and the Alaska Gastineau Mining Company donated the hatchery equipment previously used at the first trout hatchery and it was set up in a building owned by J.B. Caro and Company on Front Street in Juneau. The Alaska Electric and Power Company and the Juneau Water Company provided light and water services without charge.

About this same time the Territorial Legislature became interested in artificial propagation and passed an act which, because of irregularities, could not be funded. The Alaska Fish and Game Club continued its operations without territorial funds, except for $600 used in 1917 to finance release of trout. Another shipment of 100,000 eyed eastern brook eggs, again from Leadsville, Colorado, arrived January 18, 1918 and the subsequent fry were planted.

Sprague also experimented with salmon culture, utilizing funds provided by F.O. Burkhardt, superintendent of the Chilkat cannery of the Alaska Pacific Fisheries Company. In September, 1917, 200,000 fertilized sockeye eggs and 80,000 humpback eggs were planted in sand and gravel, at a depth of 12 to 14 inches, in a tributary spring leading from a small pond on the south side of Chilkoot Lake. No fish entered this pond as it was obstructed a short distance above its outlet into the lake. Fearing the lack of proper food for the young fish after hatching, Sprague buried the bodies of the spawned sockeye a short distance from the eggs to prevent any fungal growth spreading to the buried eggs. Sprague returned the following spring and found that a large percentage of the eggs had hatched. Sprague wrote of the results of this experiment:

> "Mr. M.J. O'Conner and Mr. Henry Roden, Mr. Martin Madsen and natives near Haines reported that on the 20th of August, 1920, thousands of Humpies were seen in the Bay and Chilkoot Lake. I do not mean to say this is positive proof, but HUMPIES were seen and this is not a Humpie district."[4]

The Alaska Fish and Game Club, undoubtedly overwhelmed by Sprague's arguments, directed the superintendent in June, 1918 to drop all trout investigations and give his full attention to salmon. Consequently, the club appealed to the fishing industry for voluntary contributions of $200 from each cannery and $50 from each saltry and mild cure plant. An anticipated $22,500 was to be collected, but only $1,100 was actually received from four packers, one of whom was

Alaska Packers Association, operator of the Fortmann hatchery. The amount was inadequate for the intended purposes and the money was refunded.

Undaunted, the club struggled along using what money it had. Superintendent Sprague and other active members served without compensation until the 1919 legislative session, when a bill was enacted along the lines of the 1917 bill. The newly created Territorial Fish Commission assumed charge of the Alaska Fish and Game Club operations on June 10, 1919. The club's hatching equipment, valued at $649.50, was turned over to the new commission for $275, an amount sufficient to settle the indebtedness of the organization. The hatchery was moved to the natatorium of the Arctic Brotherhood building, which was leased for $50 a month.

Sprague remained superintendent and swung into action. Preparations began for field work and parties installed racks in several streams on Admiralty Island and on the mainland near Juneau.[5] The hatchery was enlarged to a capacity of seven million eggs. Concrete troughs were primarily used as Sprague felt they did not collect slime and impurities as did the wooden troughs.[6] A photograph shows that some conventional wooden troughs were in use (Fig. 47). He also incubated eggs in automatic hatchery jars that previously were used for hatching shad, whitefish and grayling. This method had been declared impracticable for hatching of salmon eggs, but Sprague stated that the jars were used with great success in the Juneau hatchery.[7] Sac-fry moved to the fry tanks through a siphon.

Experimental tanks and apparatus and aquariums allowed observation of the habits of growing and adult specimens. Sprague gave tours when not actively engaged in the field and the hatchery became a popular tourist attraction. The display aquariums contained Dolly Varden, eastern brook, and cutthroat trout. One time Sprague displayed the fry which hatched from eggs from salmon that had been through the Iron Chink. One newspaper reporter wrote:

> "Here may be seen fry in every stage of deformity. Here may be seen Siamese twins, some bent in the shape of a letter S and in every conceivable shape. From these samples of fry [it would appear] that very little success will be had by using the eggs from fish that have been taken to the canneries."[8]

Sprague now had the facilities and money necessary to experiment on his theory of the superiority of planting eyes eggs over hatching eggs and holding sac-fry until mature. One of his first experiments in the autumn of 1919 was to plant eyed eggs in a controlled situation. Sprague used an unspecified small stream and dug a trench from the low bank to a depth showing seepage. Along the bottom of this trench he distributed eyed salmon eggs, then loosely filled it with gravel. The following spring when the trench was opened only fragments of the egg casings remained. Many fry of the same species as the eggs congregated near the trench.

Figure 47. Interior of Juneau Territorial hatchery, circa 1922. *Alaska Historical Library.*

Encouraged by these results, Sprague continued to plant eyed salmon eggs in many streams on Baranof, Admiralty and Douglas islands and the mainland near Juneau. Records of this period are confusing, conflicting, and incomplete (Table 28). Species planted are often not specified, often more eyed eggs were planted than eggs taken, and sources of eggs are rarely stated. It would appear that Sprague took spawn and planted fry or eyed eggs where he wanted. Sprague was undoubtedly Johnny Appleseed's counterpart in the Alaskan fisheries world.

Planting eyed eggs was not confined to streams. Sprague developed a plan to stock barren Lake Baranof with salmon. The lake empties over a large falls into Warm Springs Bay on Baranof Island and had an area calculated by Sprague to be 698 acres, with available spawning grounds, inlet streams and tributaries of ten and a half miles. He determined it was ". . . capable of holding and feeding without any cost whatsoever, one half of the entire output of young salmon (up to the age of their passing into their future home in the salt water) of the annual pack of Alaska."[9]

During the fall and winter of 1919-1920, 1,350,000 eyed sockeye eggs were placed in the lake. Official reports show that 650,000 pink eggs were deposited October 23 in gravel beds of three inlet streams flowing into the north side of the lake; 300,000 coho eggs were deposited in several places along the north shore in about six feet of water over sandy bottom, January 1, 1920; and 400,000 coho eggs were deposited at several places along the south shore in shallow water over sandy bottom on January 19, 1920.[10] Official records state

Table 28 Salmon Production, 1917-1922: Juneau Territorial Hatchery

Brood Year	Number Eggs Taken	Species	Egg Source	Distribution
1917	2,000,000	sockeye	Chilkat Lake	Eyed eggs planted, Chilkoot Lake
1917	80,000	pink	Chilkat Lake	Eyed eggs planted, Chilkoot Lake.
1919	10,540,000 3,425,000 890,000	coho chum pink	Falls, Hidden, Sprague creeks Knudson, Duck, Lund creeks (a)	Baranof Lake; Admiralty Island creeks; Juneau area streams — Knudson, Peterson, Eagle River, Duck, Salmon creek, Auke Lake tributaries; Big and Little Basket creeks; and Hadlock Lake. (b)
1920	6,460,000 5,250,000 4,664,000 670,000	coho chum pink sockeye	unknown unknown unknown unknown	Distributed as eyed eggs and fry in Lake Baranof; streams of Admiralty and Douglas islands; and mainland streams in Juneau area.
1921	unknown	unknown	unknown	Eyed pink eggs planted in Salmon, Knudson, Duck, Montana, Peterson creeks and Auke Lake in October.
1922	5,000,000 945,000 250,000 529,000 unknown	sockeye sockeye pink pink coho	Afognak Auke Lake Anan Creek unknown unknown	Distributed in Juneau area — Auke Lake tributaries, Mendenhall Bar, Auke Cove, Lemon, and Duck creeks; stream near Chas. Rudy place; Douglas Island streams; and Lake Baranof. (c)

(a) Names of some of these creeks are no longer in use and are not in Orth's Dictionary of Alaska Place Names, 1967. These are Hidden and Sprague creeks on Admiralty Island and Lund Creek on the mainland near Juneau. Collection of eggs and release of fry are listed at both Knudson and Duck creeks. Orth lists Knudson as a variation of Duck Creek (p.287).

(b) Some specifics on releases can be found in Alaska Territorial Fish Commission, report for year ended March 31, 1920.

(c) Some specifics on releases can be found in Alaska Territorial Fish Commission report for the year ended March 31, 1922, and in the Territorial Governors papers at the Alaska Archives, Juneau (particularly file 12-10-1923, letter from B.A. Rosselle to Governor Scott C. Bone, dated March 31, 1923).

that the last two plantings of eyed eggs were made through holes cut in ten inches of ice, but Sprague contradicts this in his description of the experiment, stating that 2,690,000 eyed eggs were buried before the winter coat of ice was on the lake.[11]

Observations made at the lake and its tributaries on July 6, 1920, revealed thousands of young salmon of an average length of three and one-half inches. Sprague found that some of the coho of this size migrated in October, safely descending the fifty foot outlet falls. He observed no dead fish in the pool at the base of the falls and several were taken alive in herring seines operating during the period of migration. The experiment was repeated during 1920, but details are lacking.

The only report of the return from these experiments appears in the annual report of the Territorial Fish Commission for 1927-28. Sprague states that in September, 1923 over 40,000 cohos were caught by seiners. He lists the names of four fishermen who would verify these results.[12]

Not everyone was pleased with Sprague and his techniques. In a letter from Juneau resident O.M. Harry to the Commissioner of Fisheries, Harry gives the following observations:

> "Four miles from Juneau in the fall of 1922, I observed 50,000 young fry in the egg stage dumped in the creek where I was camped. From the Juneau hatchery. They put them out of hatchery cans in three eddys. First day those three spots of minnows on the bottom was about the size of a wash bowl. There were about two dozen water ousels moved in and a lot of small trout and the next day those spots was reduced to a size of dinner plates and reduced in three days to zero. Some got away under rocks to starve.
>
> "I picked up a lot of dead ones on the salt water beach. The children and myself kept some in a tub and fed them condensed milk and bread, and I got some blood meal for them. We fed them every day, and I am safe to say that these were the only ones out of the whole bunch that did not starve to death after they dropped the eggs. We watched them nurse grass stems and try to eat each other, gradually weaken and die.
>
> "The young Alaskan salmon must be fed to the fingerling stage, then hatcheries will be a success and by cooperation of B.C. [British Columbia], Alaska and the canneries we can nauseate the world with salmon and get back our resident fishermen."[13]

Harry, obviously, was on the opposite side of the fence from Sprague, but, like Sprague, had an urge to propagate salmon. He lived for eight years at William Henry Bay (Lynn Canal) and ". . . in that time I have hatched all the different kinds of salmon of these waters with the simple means at hand, 75 this summer in a screened washtub in a spring.''[14]

Sprague continued in his enthusiasm to plant eyed eggs and fry, and it was necessary to increase the capacity of the Juneau hatchery to meet his goals. This was done in 1920 when twelve reinforced concrete troughs were added bringing the capacity to 12 million eggs.

Three years later the decision was made to close the hatchery. The last releases from the Juneau hatchery occurred on April 1, 1923, and the equipment was salvaged, packed and shipped to Ketchikan for the new hatchery there.

Eyak Lake — Cordova

Prior to the discontinuation of the Juneau hatchery, work on a territorial facility in Prince William Sound began in June, 1921 under supervision of Sprague. Two years earlier he had investigated and selected a site on a tributary, Power Creek, to Eyak Lake near Cordova (Map 13). Sprague attacked the construction with his usual enthusiasm. Land and water rights were secured, streams brushed out for seining of adult fish, roads and trails laid out, a bunk house and platform for outdoor troughs erected. A pipeline, 1,400 feet long, was completed from two springs. Diversion of water from a creek created a large pond.

About 12,000,000 sockeye eggs were taken that first year, but baskets and aerating riffle tins for the Williamson troughs failed to arrive from Juneau and, as a result, half of the eggs were buried in the sand and gravel of the natural spawning grounds in tributaries of Power Creek. The balance of the eggs were eyed in temporary troughs and were then buried in the sand and gravel of the pond. As could be expected, Sprague arranged to have some of the eggs shipped to barren lakes, courtesy of Abercombie Packing Company which donated a tender. The lakes were located in Port Wells, Cochrane Bay, Surprise Cove and in the Nellie Juan district.

All of these activities in Cordova cost more than budgeted. After three years with the Territorial Fish Commission Sprague was relieved of his duties as hatchery superintendent on December 31, 1921. It was charged that he had no authorization to begin the Cordova hatchery and proceeded without direction. He was accused of expending money extravagantly and incurring unauthorized bills.[15] A letter from the Territorial Fish Commission secretary, Charles Garfield, to the Bureau of Fisheries gives these details of Sprague's dismissal:

> ''Sprague's services were not discontinued on account of his fish cultural work, nor on account of his politics, but because he lacked executive ability, organizational and business sense. He did

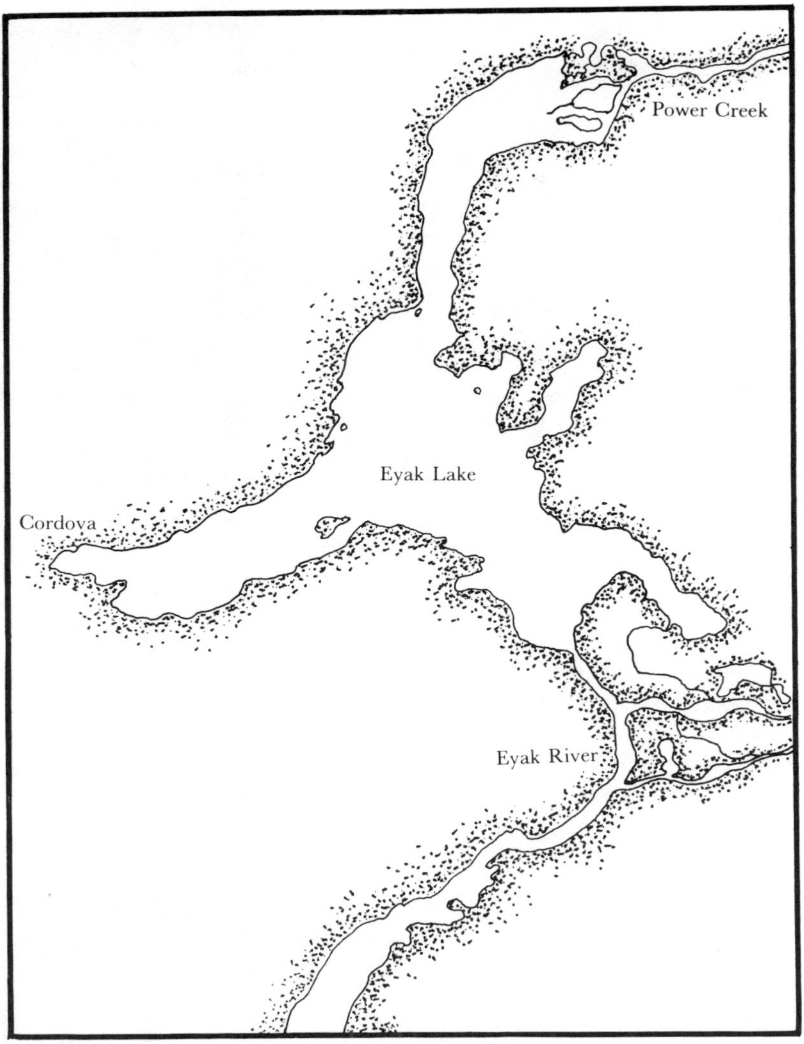

Map 13. Eyak Lake and Power Creek

> valuable work for the Territory but when the same was weighed with his mistakes, the balance was too much against him.''[16]

The Territorial Fish Commission hired Edwin Wentworth to replace Sprague. The Eyak Lake hatchery program settled into a routine enhancement of that system. Eggs were eyed in 32 troughs, 9.5 feet by 12 inches wide, on an outdoor platform. These were planted in suitable places in Eyak Lake before winter began. Eighty pairs of ripe sockeye salmon were also placed in the flooded pond which covered approximately two acres of natural spawning ground.

Figure 48. The first Cordova Territorial salmon hatchery which had outdoor troughs. *Nelson-Sprague collection, Alaska Historical Library.*

A hatchery, 35 feet by 65 feet, was erected in September, 1923 and using the old troughs had an approximate capacity of 15 million eggs. As there was no road connection with Cordova, a distance of six miles, all lumber had to be rafted and towed, then dry sledded one-half mile to the building site. Troughs were replaced in 1925 with 52 new troughs, 14 feet long and 14 inches wide and at that time it was necessary to make new wire baskets.

Figure 49. The second Territorial salmon hatchery at Cordova, five miles out of town on Eyak Lake, 1927. *Nelson-Sprague collection, Alaska Historical Library.*

In the first years of operation adult salmon were captured by seine at the mouth of Power Creek because once the fish entered the creek, they were more difficult to capture. A 1,400 foot lead was put in the lake in 1925 to aid in seining and to keep fish from entering the creek.

Eggs, taken by the incision method, were fertilized and placed in wire baskets carrying 60,000 eggs. The water circulated through the troughs and baskets at a rate of 15 gallons per minute. Average water temperature in 1923 varied from 52 °F to 38 °F.

Eggs were held past the eyed stage for the first time in the spring of 1924 when fry were fed in small ponds situated on the large flat at the head of the north arm of Eyak Lake. The second year Eyak Lake reached flood conditions, overflowed its banks, and entirely covered the flat at the head of the north arm, so that the salmon fry in their ponds were liberated before feeding had been completed. Table 29 shows the production of the hatchery.

Table 29. Sockeye Production, 1921-1925: Eyak Lake Hatchery [a]

Brood Year	Number of Eggs Taken	Date Eggs Taken	Number Eyed Eggs/Fry Released	Distribution
1921	12,000,000	July 1-unknown	6,000,000	Planted green in Spring Creek.
			5,200,000	Planted eyed near hatchery.
			800,000	Planted eyed in barren lakes in Prince William Sound.
1922	3,134,000	July 6-Aug. 1	3,078,000	Planted eyed, head of Eyak Lake.
1923	4,504,000	June 30-Aug. 10	3,882,000	Unknown distribution of swimup fry, release complete by Jan. 24, 1924.
			238,000	Fry placed in rearing ponds.
1924	5,250,000	unknown	3,242,725	Free swimming fry liberated in flood, Aug. 1925.
			1,700,000	Eyed eggs planted at Eshamy Lake.
1925	7,530,300	unknown	7,300,000	All free swimming fry held in rearing ponds and fed.

(a) 571,000 chinook eyed eggs from Washington State were incubated and 523,500 fry reared and released in August, 1925 in Eyak Lake.

Eyak Lake sockeye salmon were used, except for one shipment of 571,000 chinook eggs received from Washington State. One transfer of sockeye eggs occurred after Sprague's initial transplants. At the request of the Bureau of Fisheries, 1,700,000 green eggs were turned over to that agency to be planted in Eshamy Lake, in the western part of Prince William Sound. These eggs were eyed and planted October 7, 1924 with the aid of Bureau of Fisheries personnel. All eggs planted were well developed, as transportation problems delayed planting two or three weeks longer than necessary. Approximately 1,400,000 eyed eggs were planted five to six inches deep along the lake shore where natural seepage kept sand and gravel loose. The remainder of the eggs were planted in gravel of a stream flowing into the lake.

Shirley Baker, the Bureau's agent in Prince William Sound, examined the beds near the shore on October 28, 1924 and found some eggs had hatched, and the sac-fry were in good condition. Other beds showed that 75 percent of the eggs had died. Practically all eggs planted in the creek apparently had hatched because empty shells were found but few sac-fry.

Predator control formed an important part of the Eyak Lake program. This included the traditional elimination and exclusion of trout from the lake and the less practiced regulation of the coho salmon runs. Calvin Hazelet, the Cordova commissioner on the Territorial Fish Commission who supervised much of the Eyak Lake program believed that: ". . . heretofore we have been bothered a great deal with these silvers in the spawning ground, being a larger fish and much more aggressive they plant their eggs deeper in the gravel and destroy a great number of sockeye eggs that are almost ready to hatch."[17]

It became the practice to keep the early run of coho from mingling with the late run of sockeye. A weir was placed across the Eyak River, which enabled the Commission to count the salmon entering the lake, destroy trout and Dolly Varden, and prevent coho from entering the lake until after the sockeye were believed to have hatched. Coho were not let through the weir until September 21. Similar weirs and traps were maintained by the Territorial Fish Commision at Eshamy Lake and Robe River (near Valdez). Neither had anything to do with actual hatchery operations, but counted the salmon and destroyed the trout and Dolly Varden.

The operation of these Prince William Sound activities cost more than anticipated. The amount of money needed to finish the station, including purchase of heating and lighting plants, practically exhausted the biennial appropriations by the first of April, 1926. Activities were limited to the remaining appropriation, so the legislature was not presented with a deficit item in the 1927 session. Wentworth, superintendent of the territory's hatcheries, told the governor in May, 1926:

> "We cannot get money to operate the hatchery and weir as the people who want to operate the weir want all the silver salmon turned

over to them. No one is employed. C.C. Hazelet goes out each day and feeds fry which are all placed in outside ponds in March. He intends to move out there this summer. I personally believe that it would be better for the fry to be liberated at this time than to attempt to throw feed at them once a day."[18]

Despite pleas by Hazelet to the Cordova Chamber of Commerce and to the local cannery operators, no money was raised, and Hazelet fed the fry through the summer, and the last were liberated into the lake in November. The hatchery did not operate again. The bunkhouse, occupied by a road crew, burned to the ground during the summer of 1926. The hatchery was finally turned over to the U.S. Forest Service in July, 1927.

Seward Hatchery

The territorial facility at Seward was a short-lived operation, which lasted only three years. The hatchery was designed to be a five million egg capacity chinook salmon facility, as part of the 1923 legislative appropriation to the Territorial Fish Commission. Searches for chinook salmon from which to take eggs, however, produced negative results. The only year that chinook salmon eggs were handled was in 1924 when the hatchery received chinook eggs imported from Washington State. Very little information about the Seward hatchery has survived. Table 30 shows what little is known of hatchery production.

Table 30 Salmon Production, 1924-1926: Seward Territorial Hatchery

Brood Year	Species	Number of Eggs Taken	Number of Fry Released	Distribution
1924	chinook	1,429,000	1,387,000	Eyed eggs, Washington State. Fry fed to fingerling stage in rearing ponds. Released into Grouse Lake.
1924	sockeye	1,015,000	846,360	Eggs from Grouse Creek and Bear Lake. Fry fed in ponds and released as fingerlings into Grouse Lake.
1925	sockeye	4,460,544	4,085,727	Eggs collected, Bear Lake. Fingerlings released into Grouse Lake on July 14.
1926	sockeye	3,164,000	0	Destroyed in fire.
1926	pink	56,000	0	Destroyed in fire.

The hatchery building was located on Spring Creek at the head of Grouse Lake. It was 40 feet by 60 feet, containing 32 troughs, three deep, arranged lengthwise, and 16 troughs in pairs arranged crosswise. The egg capacity of the 48 troughs was 14,400,000 sockeye, 10,280,000 pink or 5,760,000 chinook eggs. After construction it was found that only five rows of troughs could be used during periods of low water.

The main species to be handled was sockeye and eggs were obtained from nearby Bear Lake as the Grouse Creek run was very poor. Spawning and incubation data are missing. Weirs were placed on both Bear and Grouse Lake outlets to count salmon and destroy Dolly Varden.

An area of Grouse Lake was fenced as a rearing pond and from three to four million free swimming fry were placed in this pond in late May and fed liver and condemned canned salmon. Fry were released into the lake on July 14, 1925.

Fire terminated all operations. Superintendent J.F. Starr was awakened by the barking of his dog on the night of March 9, 1927 and found the building a solid mass of flames. He barely escaped and the building with eggs and sac-fry burned to the ground. No effort was made to revive the Seward hatchery as the entire territorial program was legislated out of existence the same year.

Ketchikan Territorial Hatchery

The Juneau and Eyak Lake hatcheries were the only facilities of the Territorial Fish Commission to operate until a major reorganization of the Commission occurred in 1923. Governor Scott C. Bone, as chairman, appointed J.R. Heckman as vice-chairman. The legislative appropriation made in May was to be expended by the vice-chairman for the exclusive purpose of furthering the best interests of the Territory in the protection and the propagation of salmon.[19]

Ketchikan was selected as the place to build a permanent hatchery for the propagation of pink salmon in Southeast Alaska. This was no coincidence as Heckman was from Ketchikan and had spoken strongly many times for pink instead of sockeye salmon propagation. Heckman chose a new superintendent of hatcheries. When Sprague was dismissed at the end of 1921, the Territorial Fish Commission had asked the Bureau of Fisheries to help find another man. Edwin Wentworth, the final selection, was superintendent of the Bureau's hatchery at Afognak. He moved to Ketchikan in July, 1924, soon after the hatchery was completed.

The site selected for the hatchery was beside Ketchikan Creek only a few hundred feet from the present day Deer Mountain hatchery. A building, 40 feet by 100 feet, was equipped with 52 troughs 16 feet long, 14 inches wide and with 16 concrete troughs 10 feet long and 12 inches wide. These troughs gave a capacity of 12,180,000 pink eggs, or 17,400,000 sockeye eggs, or 6,960,000 chinook eggs. The outflow of the city's power plant provided water.

Ketchikan Creek was noted for its run of pink salmon in odd years and the fall of 1923 proved no exception. The capacity was reached with no trouble, but runs did not prove to be abundant in years after 1923 and pink eggs were procured from a number of sources (Table 31). Sockeye and chum eggs, which Wentworth used in rearing experiments, also were acquired from various sources. No details remain on adult capture methods, spawning, and incubation procedures used by the Ketchikan hatchery. All fish were held to the free swimming stage in hatchery troughs and the first year, 1923, was the only year fry were released unfed.

Like his predecessors, such as Callbreath and Sprague, Wentworth had a theory he wished to test: the value of feeding pink fry in salt water. Wentworth devised saltwater ponds in the days before floating pens. A dam of sufficient height to retain a mean depth of about four feet of water at low tide was constructed across the narrow neck of a small arm of the sea on Gravina Island.[20] Wire mesh extended above the dam, thus producing an enclosure of approximately 2,000 square feet. The only change of water in this pond was by tidal action when the enclosure was flooded for an hour or more during high tide periods. The entrance of fresh water was limited to the comparatively small amount of surface drainage from rains.

Approximately 100,000 free swimming pink fry were placed in the pond February 22, 1924. The fish soon devoured all natural food and it became necessary to feed ground liver, ground salmon eggs, and canned condemned salmon. Exact survival was difficult to determine because many small crabs in the pond made it impossible to keep an accurate check on the losses.[21]

The superintendent of the Yes Bay hatchery visited this saltwater pond twice during the spring of 1924 and reported:

> "On my arrival at this salt-water pond late in March, the fish were just beginning to take food. When I again saw them, on May 17, I was surprised to note their rapid development. There was much variation in size, ranging from 1.5 inches to 4 inches in length, but all the fish appeared to be in excellent condition. The artificial foods given were beef liver and immature salmon eggs both in the fresh state. The total loss on the lot up to the time of my second visit was reported to be 15 percent."[22]

Wentworth was very pleased with the results and he kept a photographic record of the growth of these fry as shown in Figure 50. When the fry were released in the latter part of August, 1924, the anal fins and the adipose fin were clipped from 10,000 of the fish.

Growth of fry in the Gravina pond was similar in 1925 but the size of the enclosure limited its use so it was decided to abandon this place and build at Old Gravina.[23] Although the pilings were driven in October, 1925, no record was found that a fence was constructed or that this new pond was used.

Much larger scale saltwater ponds were constructed early in 1925 with hand driven piling at Pond Bay on Duke Island and these replaced the Gravina Island locations. The fences did not hold against high water tides the latter part of March, however, and the fry were released involuntarily. Fry nonetheless remained in the locality for some time, swimming in and out of the lagoon with the tides.

Table 31. Salmon Production, 1923-1926: Ketchikan Territorial Hatchery

Brood Year	Species	Egg Source	Number of Eggs Taken	Number Fry Released	Disposition
1923	pink	Ketchikan Creek	13,167,000	9,301,000 100,000	Released as fry in vicinity of hatchery; Released as fry in Snug Harbor saltwater pond on Feb. 22; fry released late August with largest fish 7 inches. 10,000 fin marked. Also, 1,800,000 eyed eggs were shipped to Oregon.
1923	chinook	Columbia River unknown	500,000 500,000	972,655	Released in Ketchikan Creek, last fish in early August measured 4-5 inches. 5,000 fin clipped.
1924	pink	Thorne Arm River	3,477,000	2,952,000	Placed in Pond Bay saltwater ponds and released accidentally in March. Some fry placed in Snug Harbor ponds, 8,180 fin clipped.
1924	chum	Walker Cove tributaries	2,640,000	2,501,000	Ketchikan Creek as fry.
1924	chinook	Washington State	2,000,000	1,950,000	Held in ponds, some released in August; some the following spring. 10,000 fin clipped.
1924	sockeye	Yes River	13,000	11,200	Held in ponds, released April 1925, Ketchikan Creek, 4 inches long.

Permanent pilings were driven across the mouth of Pond Bay Lagoon in October, 1925, to make an approximately 15 acre saltwater lake. Fences similar to those at the Gravina site were installed (Fig. 51). Fry were first released in small netted pens to start their growth. During this period they were fed three times daily. At some point the fish were released into the large pond and fed twice daily. Growth of these fry was similar to those in previous experiments (Fig. 52).

Liver was considered the best food, but its prohibitive cost limited its use to starting fry to feed. Once fry were feeding well, salmon spawn were used. These were taken from the Iron Chink, the unripe milt being taken as well as the roe, and placed in cold storage. This food was cooked and ground for initial feedings. Later, when fingerlings were larger, they would take fresh eggs and milt.

Table 31 — Continued

Brood Year	Species	Egg Source	Number of Eggs Taken	Number Fry Released	Disposition
1925	pink	Ward Cove	8,028,000	4,250,000	Released in Pond Bay ponds, released end of July.
		Ketchikan Creek	3,387,000	5,497,000	Released in March in Ketchikan Creek.
1925	chum	Walker Cove	40,000	200,000	Released in Pond Bay feeding ponds.
	chum	Crab Bay	280,000	65,000	Released in Ketchikan Creek.
1925	sockeye	Ward Cove	165,000	161,000	Held in hatchery ponds until March 1, 1926.
	chinook	Kalama, Washington	2,000,000	1,952,000	Held in hatchery pond, released at 6 inches in Ketchikan Creek, Sept-Nov., 1926.
1926	sockeye	Buschmann Creek	1,320,000	2,625,000	Released as fingerlings in Ketchikan Creek.
	sockeye	Ward Cove	300,000		
	sockeye	Yes River	1,717,760	400,000	Held for overwinter feeding.
1926	pink	Ward Cove	150,000	574,000	Released into Ketchikan Creek.
	pink	Lucky Cove	1,510,000	950,000	Released in Pond Bay ponds.
1926	chinook	Washington State	2,000,000	1,819,000	Released as fingerlings into Ketchikan Creek, starting June, 1927.
1926	chum	Ward Cove	451,000	225,000	Released into Ketchikan Creek.

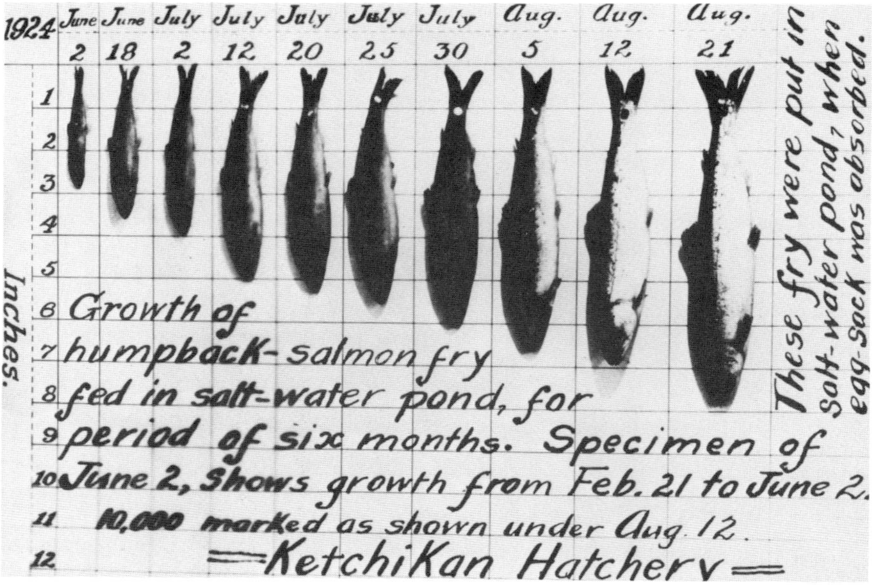

Figure 50. Growth of pink fry fed in saltwater pond for six months, Ketchikan Territorial hatchery, 1924. *Northwest collection, University of Washington.*

Figure 51. Saltwater rearing pond used for pink and chum fry at Pond Bay by Ketchikan Territorial hatchery, 1925-1926. *National Archives.*

Figure 52. Specimens of pink fingerlings fed in saltwater ponds, Pond Bay by Ketchikan Territorial hatchery, 1926. *National Archives.*

Chum salmon were also fed at the Pond Bay ponds in 1926 and 1927. They made "a very satisfactory growth, much more uniform than the humpbacks and they seemed to be more hardy," Wentworth reported.[24] Additionally some chum fry were released into Ketchikan Creek.

Another of Wentworth's programs centered on chinook salmon. Eyed pink eggs were exchanged with the state of Oregon for 500,000 Columbia River chinook eyed eggs in 1923. The Bureau of Fisheries sent an equal number from an undetermined source. This began a four year program of releases of this species into Ketchikan Creek. In what is today the city park, Wentworth built freshwater ponds, to hold the chinook fry after they had started feeding in the hatchery troughs. As the fish increased in size it was necessary to liberate some to avoid crowding, as two million fry taxed the pond's capacity. The screens in the ponds were usually removed in August and the fry allowed to work out if they were inclined. By this time the larger fish were four inches in length. Many of the fish stayed in the ponds the entire winter and moved into the creek where they still could be seen in April. Wentworth had plans and estimates to construct suitable ponds to feed all chinook fry. These plans were sent to the governor for consideration and were to be submitted to the 1927 legislature, but the territorial hatcheries were abolished that year.

Sockeye salmon also interested Wentworth. In fact, coho salmon was the only salmon species with which he did not experiment. He obtained 13,000 sockeye eggs in 1924 from the Yes Bay hatchery and reared the fry all winter, liberating them the first week in April at an average length of four inches. Wentworth said of this project: "The feeding of these fish through the winter demonstrates that it is possible to carry them in good shape and liberate them at a time when the wild red salmon fingerlings are migrating from fresh to salt water."[25] The next year Wentworth found his sockeye fingerlings were larger, though six months younger, than those of the previous year.

Wentworth was aware that no run of sockeye salmon existed in Ketchikan Creek. Therefore, a return of these fish would demonstrate the possibilities of sockeye salmon culture when feeding was carried on for 18 months under what he considered proper conditions. He hoped, since Ketchikan Creek is a lake fed stream, that some adult salmon would return to prove his theory. Future plans included a feeding pond for sockeye fry on a creek above Ward Lake, where some of the spawn had been taken.

All of Wentworth's programs came to a halt in 1927 with the elimination of the Territorial Fish Commission. He was able to feed the fish on hand until they normally would be released, partially with an appropriation pushed through by the Nome legislator Ralph Loman and partially with contributions by John Koel, a Ketchikan businessman.

It was unfortunate that the program ceased at a time when Wentworth expected results and planned for evaluations. Wentworth marked fingerlings with varying fin clips during the four years the hatchery operated. Each of the chinook brood years had different clips. The pink fry at Pond Bay were marked with different combinations of ventral, dorsal and adipose fin clips to differentiate years of release. A poster showing the species and the fin marks asked that fishermen lend support by reporting catches of these marked salmon to the Territorial Fish Commission or the U.S. Bureau of Fisheries in Juneau. A reward of $1 was offered one year for marked pink salmon. A report of one pink salmon was received from the 8,000 fingerlings marked in 1925 at the Gravina Island pond.[26] No other records were found which indicated if these signs were posted or if the fishermen responded.

Wentworth checked the area of Pond Bay Lagoon in 1926, but found no large numbers of fish in the locality. If there had been, these would have been a consequence of the fry which escaped the lagoon almost immediately after release and had been fed very little, if at all.

Once the Territorial Fish Commission program ceased, no records appear to have been kept. J.R. Heckman continued his interest by checking returns to Ketchikan Creek. He reported a few chinook and sockeye salmon above the falls in late September, 1930.[27] The next year Heckman reported "Just to the rear of the hatchery may be seen a number of sockeye salmon right at its very door."[28] A Bureau of Fisheries employee counted over 400 sockeye salmon which ascended a fish ladder at the falls in Ketchikan Creek in 1931. Before the hatchery was established, it was said, not a single sockeye or chinook salmon had been seen in the creek.[29]

Emery Tobin, a long-time Ketchikan resident wrote of the chinook salmon returns:

> "I was a long time good friend of Ned Wentworth.... He had been experimenting with the hatching of king salmon.... with some success in as much as some king salmon identified by having one small fin removed came back to Ketchikan Creek. I believe these had been reared at the City Park hatchery."[30]

Money was appropriated for one more year after abolition of the Territorial Fish Commission so a report on the condition of the fisheries could be made. The men who had first been intensively involved with the Territorial Fish Commission were gone and A.J. Sprague was able to obtain the job of checking on the conditions of the fisheries and to do small amounts of fish culture work.

One of Sprague's first acts was to recommend complete closure of the Ketchikan hatchery because of the rearing ponds. Holding fry in ponds was something that Sprague still could not abide. The hatchery was closed, but the Ketchikan Commercial Club, by petition, requested the Territorial legislature to restore use of the hatchery in 1929. This was not done and the hatchery was deeded to the city of Ketchikan in 1931, but was not used for that purpose.

Sprague completed his investigation into the fisheries during the fall of 1927. His desire to plant eggs did not diminish and he planted a few pink eggs (Table 2). In his year-end report, he suggested:

> "Juneau should have one [hatchery] to take care of the rehabilitation at or near Montana Creek and instead of hatching out the eggs and distributing the fish, to develop the eggs to the eyed stage and distribute them in barren lakes and streams as has been successfully done. This does away with prohibitive cost of feeding.... "[31]

The Territory had come full circle on artificial propagation theories. It chose not to continue its involvement, but when it did renew its interest in the 1950s, it chose to stock barren lakes.

Notes

1. A.J. Sprague, *The Alaska Salmon and Their Practical Propagation* (Juneau: Daily Empire Print, 1921), p. 15

2. Ibid, p. 16.

3. Ibid.

4. Ibid, p. 7.

5. Names of some of the creeks are no longer in use and are not in Donald J. Orth's *Dictionary of Alaska Place Names,* 1967. These are Hidden and Sprague creeks on Admiralty Island and Lund Creek on the mainland near Juneau. Collection of eggs and release of fry are listed at both Knudson and Duck creeks. Orth lists Knudson as a variation of the name Duck Creek (p. 287). Two other egg sources are Peterson and Falls Creek, both identified in Orth.

6. *Alaska Dispatch,* Juneau, February 13, 1920.

7. Ibid.

8. Ibid.

9. Sprague, p. 8.

10. Alaska Territorial Fish Commission, *Report of the Alaska Territorial Fish Commission for the Year Ended March 31, 1920* (Juneau, 1920), p. 6.

11. Sprague, p. 8.

12. A.J. Sprague, *Report of the Territorial Fish Commissioner to Territorial Legislature on H.C.R. No. 14, Session Laws,* 1927-28 (Juneau, 1928), pp. 10-11.

13 .Letter from O.M. Harry to the Commissioner of Fisheries dated March 1, 1928. File "Miscellaneous Requests-Fisheries." General Records, Division of Alaska Fisheries, RG 22, NA.

14. Ibid.

15. Letter from Charles D. Garfield, Secretary of the Alaska Territorial Fish Commission to Commissioner of Fisheries dated February 3, 1922, File "Territorial Fish Commission, 1918-1929," General Records, Division of Alaska Fisheries, RG 22, NA.

16. Ibid.

17. Letter from Calvin C. Hazelet to Governor Scott C. Bone dated November 26, 1924, File "12-10-1924," Records of the Office of the Territorial Governors, Alaska State Archives, Juneau.

18. Letter from Edwin Wentworth to Governor George A. Parks dated May 21, 1926, File "12-10-1926," Records of the Office of the Territorial Governors, Alaska State Archives, Juneau.

19. Alaska, *Laws of Alaska, 1923* (Juneau, 1923), pp. 182-183.

20. This harbor was called Snug Harbor by Wentworth and does not appear in Orth's *Dictionary of Alaska Place Names,* 1967.

21. Letter from Edwin Wentworth to Governor Scott C. Bone dated February 25, 1925, File "12-10-1925," Records of the Office of the Territorial Governors, Alaska State Archives, Juneau.

22. Glen C. Leach. *Propagation and Distribution of Food Fish, Fiscal Year, 1924,* Bureau of Fisheries Document No. 978 (Washington: Government Printing Office, 1925).

23. Old Gravina was located in a small bight which has been obliterated by construction of the Ketchikan International Airport.

24. Letter from Edwin Wentworth to Governor George A. Parks dated November 5, 1926, File "12-10-1926," Records of the Office of the Territorial Governors, Alaska State Archives, Juneau.

25. Letter from Edwin Wentworth to Governor George A. Parks dated April 1, 1926, File "12-10-1926," Records of the Office of the Territorial Governors, Alaska State Archives, Juneau.

26. Alaska. *Report of the Governor of Alaska to the Secretary of the Interior, 1926* (Juneau, 1926), p. 42.

27. *Ketchikan, Alaska Chronicle,* September 23, 1930.

28. *Ketchikan, Alaska Chronicle,* September 10, 1931.

29. Ibid.

30. Letter from Emery Tobin to Patricia Roppel dated October 21, 1973.

31. Sprague, *Report of Territorial Fish Commissioner of Alaska,* p. 17.

VII. Renewal of Interest

Little interest was demonstrated in salmon propagation for the fifteen years after the closure of the last cannery-operated hatchery in 1936. A policy of fishing regulation, instead of artificial propagation, was the method adopted to keep Alaska's fisheries at satisfactory levels. This continued until 1951 when the Territory of Alaska entered once more into an active enhancement and research program.

Fish culture was not completely forgotten during the period of regulation. A few individuals queried the U.S. Fish and Wildlife Service, the agency responsible for Alaska's fisheries after 1940. The director made this typical response in 1940 to an inquiry into opportunities for private fish culture:

> "Careful study of these [past] operations, however, disclosed that natural propagation is more effective than artificial propagation in Alaska and no hatcheries are operated now in Territorial waters. Private hatcheries, therefore, are not encouraged.
>
> "Under existing laws the construction and operation of private salmon hatcheries cannot be prohibited. Such operation, however, if for commercial purposes, would be subject to all the prohibitions applicable to commercial fishing, including restrictions on the time and place where salmon fishing can be conducted. The carcasses of the salmon also would have to be used to avoid penalty for wanton waste of food fish."[1]

Little Port Walter Experimental Hatchery

Only once during that fifteen year span from 1936 to 1951 was operation of a hatchery pursued. This experimental pink salmon hatchery was part of a research program started by the Bureau of Fisheries at Little Port Walter in the south Chatham Strait area of lower Baranof Island. At the biological station and two-way salmon counting weir, intensive pink salmon studies began in the early 1940s. Experiments researched the type and amount of spawning gravel required per pair of pink salmon, their sex ratio and egg production, and factors affecting the freshwater survival of the pink salmon eggs and fry. A fry marking program gave additional information on homing instincts and degrees of straying.

It was into this background of investigations into the natural run that an experimental pink salmon hatchery was introduced. The experiment was conceived by Frederic F. Fish, chief of the Western Fish Cultural Investigations, a part of the Branch of Fishery Biology of the U.S. Fish and Wildlife Service, and by Seaton Thompson, Chief of the Alaska Division. It attempted to measure the impact of artificial pink salmon propagation on the natural run of pink salmon to Sashin Creek.

Frederic Fish was in charge, and the fish culturist attached to the project was Roger E. Burrows, then in charge of the Salmon Cultural Laboratory at Leavenworth, Washington. The pair were assisted in Alaska by James Drewry and J.M. Olson of the Alaska Fisheries Investigation staff. The experiment called for artificial production of pink fry which would be marked at the time of their normal seaward migration in March, 1948. Naturally produced pink fry from Sashin Creek would be marked with a different fin clip so that with direct comparison of artificial and naturally propagated fry the difference, if any, in the marine survival rate could be analyzed. If manpower, funds, and circumstances permitted, a second group of artificially propagated fry would be marked with a still different fin clip and released in the stream entering Lover's Cove, Big Port Walter, to determine whether such relocated fish would return to the site of incubation or to streams in which liberated. A cabin by the two-way weir at Little Port Walter was transformed into a hatchery during the summer of 1947 by Harlan E. Johnson. Frederic Fish, Burrows and their assistants selected a donor stream, East Creek, in Tebenkof Bay across Chatham Strait from the research station. Securing sufficient ripe females at one time proved difficult during September. On the first day spawn was taken an average of 60 salmon were handled for each ripe female located. Frederic Fish recommended in his progress report that beach seining was an unsatisfactory way to secure large numbers of sexually mature female salmon at one time if the East Creek experience was typical of Southeast Alaska.[2] The crew secured only 178,820 eggs. A storm forced the crew to wait 30 hours with 50,000 of these fertilized eggs in a full, sealed 10-gallon milk can. Eggs in this lot had a higher mortality rate than lots arriving at the hatchery in shorter time.

Frederic Fish and Burrows left Little Port Walter after placing the eggs in the hatchery. The majority of the eggs were lost because an over-conscientious, inexperienced employee removed dead eggs throughout the critical period of development despite instructions to the contrary.[3] Approximately 19,000 to 20,000 fry resulted. These fry and naturally produced fry from Sashin Creek were marked and released. Frederic Fish, disappointed in the difficulties encountered, stated:

> "The decision whether pink salmon propagation will be resumed at Little Port Walter next year [1948] should be influenced by the relative overall importance of the experiment as well as the needs for rehabilitating the pink salmon population in Sashin Creek during the fall of 1948."[4]

The low egg survival and limited fry release invalidated this experiment and it was not continued.[5] The records showing weir counts and fin marks collected from the experiment were missing when a summary of research at Little Port Walter was written in 1956.[6]

Territorial Alaska Department of Fisheries

Two years after the Little Port Walter experiment, there was a renewed interest by the Territory in its fishery resource. The Territorial legislature created the Department of Fisheries in 1949 to assist in conservation and perpetuation of this important resource; to promote more resident ownership, management and control; and to cooperate with the U.S. Fish and Wildlife Service, which had jurisdiction over Alaska's fisheries. Most of the enhancement projects undertaken by the newly created department were not oriented toward hatchery production of salmon. During the ten years prior to statehood, only two hatcheries were built, one at Ketchikan — the Deer Mountain hatchery — and the other at Kitoi. The latter was part of a research station whose main purpose was not to produce quantities of fish in a hatchery environment.

Immediately after creation of the department, it was determined that the initial projects would concentrate upon the troll fishery for chinook and coho salmon. As a first step in investigation of the potential chinook production of Southeast Alaska streams a shipment of 35,000 eyed chinook salmon eggs was obtained from the Washington Department of Fisheries and arrived in Alaska on December 20, 1950. It was hoped that runs of fall chinook could be started in small streams of the territory which were unsuitable for spring run chinook. These eyed eggs were transplanted in two streams in the Mendenhall Valley near Juneau where accurate checks could be made upon their development.

The second shipment of eyed fall chinook eggs arrived on January 5, 1952 from the Soos Creek stock near Seattle, Washington, probably the same source as the first shipment. The 66,000 eyed eggs were planted in Montana Creek, a tributary of the Mendenhall River. A basin was first dug 20 inches deep and three feet in diameter in the stream bed. Then the downstream end of a five foot length of two inch tubing was placed in the bottom of the basin so that there was a gradual slope to the tube providing a rapid flow of water through the tube. Coarse gravel was heaped over the lower end of the tube until it filled the basin. The eggs were poured into the upper end of the tube, which was placed at the surface of the stream, and the flow of the stream distributed the eggs into the spaces of the gravel.

The third shipment of 60,000 eyed eggs arrived December 17, 1952, and a fourth shipment of an undetermined number of eggs on December 13, 1953. Both shipments were placed in Spring and Montana Creeks. No report has been found of returns or attempts to locate returns from these four transplants.

In addition to the eyed chinook salmon egg program near Juneau, sockeye salmon spawn was taken from the Lake Creek run (a tributary to Auke Lake). These eggs were eyed and in October, 1951 planted in the inlet and outlet creeks of Dredge Lake and in Duck Creek in the Mendenhall Valley. This program was abandoned after the first year.

Planting of eyed eggs, of course, was not a new practice in Alaska. A.J. Sprague had been the chief advocate of this method of salmon enhancement and his influence was still strong when the Territory again began fish culture work in the 1950s. C.L. Anderson, the director of the newly created Territorial Department of Fisheries hired Sprague to operate an egg eyeing station which was located at 236 Irvine Street in Juneau, using water from Gold Creek. This hatchery operated until Sprague's death in 1952. But before his death Sprague had influenced the course of the Territory's fish culture program: eyed eggs would be planted and barren lakes would be brought into production.

Many lakes and streams are not accessible to salmon because of natural barriers or obstructions. The removal of these or construction of fishways around them would open up new spawning areas. Because of uncertainty whether salmon would naturally select these new areas, it was decided to stock eyed eggs to start runs so a return would be certain.

The first experiment began in the Pauls Lake System on the northeast coast of Afgonak Island, a system which includes three lakes, Pauls, Laura and Gretchen, with connecting streams offering additional areas for spawning. Only Pauls Lake supported runs of salmon — sockeye, pink, and coho — and a barrier of four falls between lakes prevented ascent by salmon.

A substantial run of sockeye was located in the adjacent Perenosa (Portage) Lake system and mature salmon were seined in the mouth of the spawning stream. Standard spawning procedures were used and the eggs were drained and packed in moss-padded baskets for transport. Eighty-four thousand green eggs, transported to Gretchen Creek, were immediately planted. The remainder of the eggs, 22,000, were transported to a temporary egg incubator in Port Williams. Additionally 107,000 eggs from Karluk Lake were eyed and both lots of eggs were taken to upper Gretchen Creek, making a total of 213,000 green and eyed eggs planted in August and September, 1951.

The following spring sockeye salmon fry were seined from the mouth of this creek, indicating the successful introduction of salmon into the system. As a result, a temporary egg eyeing station was constructed on a small tributary of Perenosa (Portage) Creek, one-half mile upstream from the head of Discovery Bay. The station, with a capacity of 500,000 eggs, consisted of a lean-to shelter with two 16 foot standard Williamson troughs and one vertical drip incubator. Sockeye eggs taken from Southeast Creek, a tributary of Perenosa (Portage) Lake, were transported to the eyeing station. During incubation the eggs were treated with malachite green hydrochloride approximately twice a week for half an hour intervals, then flushed clean of the solution. No evidence of fungus was observed during or after the incubation period. A total of 488,000 sockeye eggs were eyed in 1954 with a mortality rate of 2.5 percent.

Eyed eggs were initially planted by use of the five foot tube, previously used in the department's chinook program in Juneau. A new method was used after

1955, similar to one developed by the International Pacific Salmon Commission. Trenches were dug parallel to the current in the stream gravel in suitable locations. A cylinder, 24 inches in diameter and 36 inches high, was placed at the upstream end of the trench with another one downstream from it. A few large rocks, six inches in diameter, were placed in the upstream cylinder, and from 1,000 to 8,000 eggs poured into the cylinder. When the eggs had settled into the interspaces of the rocks, gravel from the stream bed was shoveled into the cylinder until slightly higher than the surrounding stream bed. Then the first cylinder was placed in the trench downstream from the second cylinder. This action left a slight mound of gravel over the eggs. This system seemed superior to the tube method as it was quicker and gave better distribution of eggs throughout the gravel.[7]

The last eyed eggs were planted in Gretchen Creek in 1955 when adult sockeye salmon had begun to return. It was hoped the run would be self-sustaining. By this time the department had embarked on a similar project at Fraser Lake, located on the southern end of Kodiak Island. Its outlet drained into Olga Bay over a 33 foot falls about eight-tenths of a mile below the lake. Again the department used an incubation station to eye eggs, this time in an inoperative cannery at Olga Bay. The available water was more than adequate for the four sixteen foot troughs (Fig. 53). Eggs eyed out in approximately 30 days.

Figure 53. Olga Bay eyeing station, Kodiak Island, 1955. *Alaska Department of Fisheries.*

From 1951 until 1955 the source of eggs for the Fraser Lake project was the Karluk River. A new source of sockeye eggs became necessary in 1956. East Creek (Crystal Creek), a tributary of Red Lake, a sockeye system adjacent to the Fraser drainage, provided the necessary eggs. This site was selected because the migration from salt water to the spawning grounds was considered a comparable distance to that involved in the ascent of the chum salmon in the Fraser River. That year, 1956, 268 females and 208 male sockeye were spawned and 577,719 eggs taken. Water temperatures at the eyeing station ranged from 45 °F to 50 °F and 500,444 eyed eggs were planted, yielding a 13.3 percent mortality rate to that stage. The last year prior to statehood in which eyed sockeye eggs were planted at Fraser Lake was 1956. Returns continued to increase over the years after statehood.[8]

Kitoi Bay Research Station

Introduction of sockeye salmon to Fraser Lake and Pauls Basin progressed, but the Department of Fisheries was hampered by the lack of scientific knowledge on how to, how much to, what to, and where to plant barren lakes. Realizing the need to answer these questions, the Alaska Fisheries Board secured funds from the Alaska Legislature to provide research facilities. The site chosen was on the southeast coast of Afognak Island at Kitoi Bay, an area where numerous lakes provided opportunity for study of sockeye and coho salmon production.

Construction of the Kitoi Bay Research Station commenced in the fall of 1953 and was completed in 1954. Late that year a hatchery, 84 feet by 24 feet, was built adjoining the laboratory and put into operation after the installation of a pipeline in 1955. The principal objective of the research station in the period prior to statehood was to develop techniques for introduction of sockeye into lakes which were not utilized by anadromous fish. The hatchery provided fry for various experiments in stocking these lakes. Some research was also conducted on hatchery techniques, such as on the use of Roberson stacked trays, on non-circulating egg trays, on an immersed egg box suspended below the surface of the lake, and on ripening adult sockeye.[9]

There is little data available on the Kitoi station. Most of the original information on hatchery operations, egg, and fry plants in the Kodiak area was never documented except in field notes, was simply lost or misplaced, or was lost during the destruction of the hatchery in the 1964 earthquake and subsequent tsunami.

Deer Mountain Hatchery-Ketchikan

Emphasis on stocking barren lakes was expanded to the Ketchikan area. Enhancement of the Old Frank's Lake system on Prince of Wales Island was chosen as the first project in 1952. Four lakes, a total of 872 acres, with uneven shorelines and numerous shallow bays offered potential rearing area for young salmon. To test the potential, 60,000 green sockeye eggs were planted in October, 1952 and 77,000 eggs were planted in 1953. Also placed in the gravel of the streams were 35,000 sockeye eggs which had been eyed at a small station on a tributary of Ward Lake near Ketchikan. The project was abandoned after two years and the only analysis of its success appears to have been observations made in 1956-57. Actual count of sockeye seen jumping at the falls to the system revealed 25 jumps observed in a period of ten minutes in August, 1957.

Surveys of other promising systems took place during the period of the Old Frank's Lake system experiment and in 1953 the Bakewell Lake system took priority. This lake drains into an arm adjoining Wilson Arm to form Smeaton Bay south of Ketchikan. A barrier falls lies 200 yards below the lake in the half-mile long outlet stream. The system offered possibilities of opening a large lake area for natural production of salmon.

Studies proved the feasibility of the plan, and the first sockeye eggs were brought in during the fall of 1954. The department utilized the run at Buschmann Creek, the stream beside the abandoned Quadra hatchery, as the source of these eggs. Eighteen years after the hatchery closure the run was sufficiently strong to provide a constant source of spawn for the department for the six years before statehood. Most of the eggs were water hardened in sealed one gallon thermos jugs, then shipped to the incubation station in Ketchikan, a lapse of three or four hours. During August and September, 1954, 100,000 green and 200,000 eyed sockeye eggs were placed in the inlet streams of Bakewell Lake. Examination of test cartridges filled with fertilized eggs revealed very low mortality rates after one month.

The Bakewell Lake project was greatly enhanced in 1955 when the Deer Mountain hatchery started production. Eyeing of eggs had been accomplished under less than ideal conditions in the local U.S. Fish and Wildlife hangar. With the Deer Mountain hatchery, it was possible to plant fry in the lake, and this was carried out every year until statehood (Table 32). The successes and problems of the Bakewell Lake project continued after Alaska became a state and gained control of its fisheries.

The Deer Mountain hatchery was the key to the department's program in the Ketchikan area. It was built by the Ketchikan King Salmon Derby Committee, sponsored by the Ketchikan Chamber of Commerce, and outfitted by the Alaska Department of Fisheries. Ground was broken on August 2, 1954 on a site only a few hundred feet from the site of the Territorial Fish Commission hatchery operated in the 1920s. On April 22, 1955, Governor B. Frank Heintzleman delivered the dedication speech.

A cedar log building, 40 feet by 58 feet, contained 48 Williamson troughs and had the rearing capacity for one and one-quarter million fry. The head

Table 32 Releases of Salmon Fry from Deer Mountain Hatchery in Bakewell Lake, 1955-1960

Release Date	Species	Number of Fry	Number Per Pound	Egg Source
7/9/55	coho	71,100	900	Soos Creek, Wn.
7/9/55	coho	15,000	700	Soos Creek, Wn.
9/4/55	sockeye	49,700	710	Buschmann Creek
6/13-10/1/56	sockeye	301,200	150-1,400	Buschmann Creek
5/19-5/22/57	sockeye	220,000	2,440	Buschmann Creek
6/14/57	sockeye	27,300	1,560	Buschmann Creek
8/14/57	sockeye	44,000	846	Buschmann Creek
10/3/57	sockeye	10,800	271	Buschmann Creek
11/5-11/6/57	sockeye	21,000	179	Buschmann Creek
1958	sockeye	11,500	...	Buschmann Creek
8/20-8/21/59(a)	sockeye	67,200	1,426	Buschmann Creek

(a) Fishpass completed 8/24/59. Sockeye spawners observed in main tributary, Bakewell Lake.

trough, which fed the 24 series of two troughs each (end to end thus giving double use of water), was built of reinforced concrete and was 40 feet long, 2 feet deep and 2 feet wide with a sloping bottom which drained toward a six inch standpipe. An eight inch gate valve controlled the outlet flow at the headbox. Water flowed 425 feet underground from the municipal powerhouse tailrace. Ponds in the city park adjacent to the hatchery were adapted for use in rearing fingerlings. These ponds had originally been designed for the territorial hatchery and had been incorporated into a city park. A concrete fish ladder leading to a concrete holding tank to capture and contain returning salmon was added in 1957.

The main objective of the hatchery was to provide salmon fry for lake stocking programs. It was believed that fry gave a much greater flexibility to the program than eyed eggs because release of fry could be timed to conform with the maximum production of plankton in the watersheds.

Another objective was to provide chinook and coho salmon for the troll and sport fisheries. Attempts to locate an Alaskan supply of chinook salmon eggs began in 1955, but preliminary surveys made in the Unuk and Chickamin systems near Ketchikan showed few of this species available. It was decided to again attempt the establishment of a fall run of chinook to short streams in Southeast Alaska. In 1955, 118,000 eggs were attained from Soos Creek, Washington. Yearly shipments were received at Deer Mountain hatchery for

Table 33 Releases of Salmon Fry from Deer Mountain Hatchery into Ketchikan Creek, 1956-1960

Release Date	Number of Fry	Number per Pound	Egg Source
		CHINOOK SALMON	
9/17-11/9/59	89,500	75-260	Soos Creek, Wn.
8/56/to 3/57	...	...	Soos Creek, Wn.
6/57/to 3/58	83,000	332-415	Soos Creek, Wn.
2/23, 25, 29 to 3/31/60	38,436	79	Soos Creek, Wn.
		COHO SALMON	
1/9/56	53,300 (a)	143	Soos Creek, Wn.
9/9/57	24,000 (b)	191	Reflection Lake and Ward Cove
10/2/57	10,000 (b)	111	Reflection Lake and Ward Cove
11/13/57	7,000 (b)	81	Reflection Lake and Ward Cove
8/4/59	2,500	645	Ketchikan Creek
4/19/60	16,406	65	Unknown
9/10/60	63,442	150	Unknown

(a) 2,200 marked at release by removal of adipose fin
(b) 12,500 from the Reflection Lake stock released in 1957 were marked by removal of the adipose and left ventral fins.

four years. The eyed eggs were hatched, the fry fed for a period of time in the hatchery ponds, and fry migration into Ketchikan Creek appears to have been volitional beginning in August and ending the following May (Table 33).

Results of this attempt to develop a run of chinook to Ketchikan Creek prior to 1960 show one jack returned to the hatchery trap in 1957. The following year 80 three year old jacks ranging in size from 2.5 to 10.5 pounds came back to the hatchery trap. The run started on September 18 and continued through October 28. The records for 1959 have not been located, but in 1960 58 adult chinook returned, ten of which were females. From this return 47,500 eggs were taken. After statehood the chinook program continued until 1967.

The other thrust of the Deer Mountain hatchery was coho salmon. Location of adequate numbers of this species in Alaska also proved difficult. In order to start the program with sufficient numbers of fry to release elsewhere than in Ketchikan Creek, a transfer from Washington was made. On January 20, 1955, 315,000 eyed coho eggs of the 1954 brood year taken from Soos Creek stock at the Green River hatchery near Auburn, were placed in the Deer Mountain hatchery. Also in the hatchery were 18,000 coho eggs from Buschmann Creek stock.

The Buschmann Creek coho eggs hatched during the first two weeks in March and the Soos Creek eggs reached a peak on March 18. Feeding of all coho fry began on June 1 and continued for at least one month. That year the fish lived on condemned worm-infested beef and hog liver which could be obtained at nominal prices. Salmon viscera was also part of the diet, much of it contributed by local trollers and the Ketchikan Cold Storage. All food was stored free by New England Fish Company. To start the fry, 50 percent beef and 50 percent hog livers were fed. After fish were feeding well, the diet was changed to one-third each of hog and beef liver and salmon viscera. Distiller's solubles provided a mineral and vitamin supplement.

Alaskan sources for coho eggs were relied upon after this initial, imported shipment. A few eggs were taken at Reflection Lake inlet stream in 1955, and from fish seined from the base of Ward Cove dam in 1956.

Releases of the fry from Soos Creek, Reflection Lake, and Ward Creek egg sources established a coho run in Ketchikan Creek (Table 33). This was verified when 65 adult coho from the Soos Creek stock returned to the hatchery via the fishway in 1957. Two were marked by removal of the adipose fin (Fig. 54). In

Figure 54. Marked coho salmon returned to Ketchikan Creek, Deer Mountain hatchery, 1957. *Alaska Department of Fish and Game.*

addition, two marked jack coho of the Reflection Lake stock returned. Four other jacks entered the trap, but escaped before identification.

The next year the Reflection Lake release produced 228 marked adults which were caught in the holding pen between July 14 and November 29. Two marked coho were reported in the troll fishery that year. One hundred thirty-nine unmarked coho also returned in 1958 and it was reasoned these were four year olds from the Soos Creek fish.[10] An analysis of scales was supposed to have been made, but no results were located. From the returning adults 284,000 eggs were taken.

The coho, particularly the abundance of Soos Creek origin fry, were planted in many streams and lakes in the vicinity of Ketchikan in an attempt to establish runs to barren lakes and streams (Map 14 and Table 34). Fry were transported in airplanes in square five gallon cans aerated by a six-volt pump to drive air through plastic tubing to air stones in the bottom of the cans. Five pounds of fish were shipped per can and sodium amytal (one-half gram per gallon) reduced the metabolism of the fish. This increased by approximately 100 percent the number of fry it was possible to carry per can. Three to four pounds of ice per can reduced water temperature to 33 °F to 35 °F. Water was then tempered before release of fry. Time in transit was less than an hour. If analyses of the success of these releases were made prior to statehood, the reports have not been located.

Some coho releases were in connection with scientifically designed research. After the success of the Kitoi Bay research the department cleared Princess Lake near Ketchikan of resident fish by treatment with rotenone prior to release of coho and sockeye salmon fry (Table 35).[11]

The object of the study at Princess Lake was to determine the extent coho were competitors of sockeye salmon and to find if more pounds of desirable fish could be raised by mixing the species or keeping them separate. Fry were released again in 1959 and the downstream migrants trapped as they left the system.[12]

Resident fish were cleared from Crane and June lakes, which drain into Ideal Cove by Petersburg, Lake Arrowhead in East Behm Canal near Ketchikan, and a series of lakes draining into Tsa Cove in George Inlet near Ketchikan (Map 14). A program was started to plant these lakes with salmon and this was carried out in 1958 and 1959 (Table 34).

The Territorial Department of Fisheries' research projects, designed to increase knowledge necessary for salmon enhancement programs, were continued until the end of December 1959. At that time there began a new era in fisheries. Alaska was now a state and had finally been granted the much desired ability to manage its own fisheries.

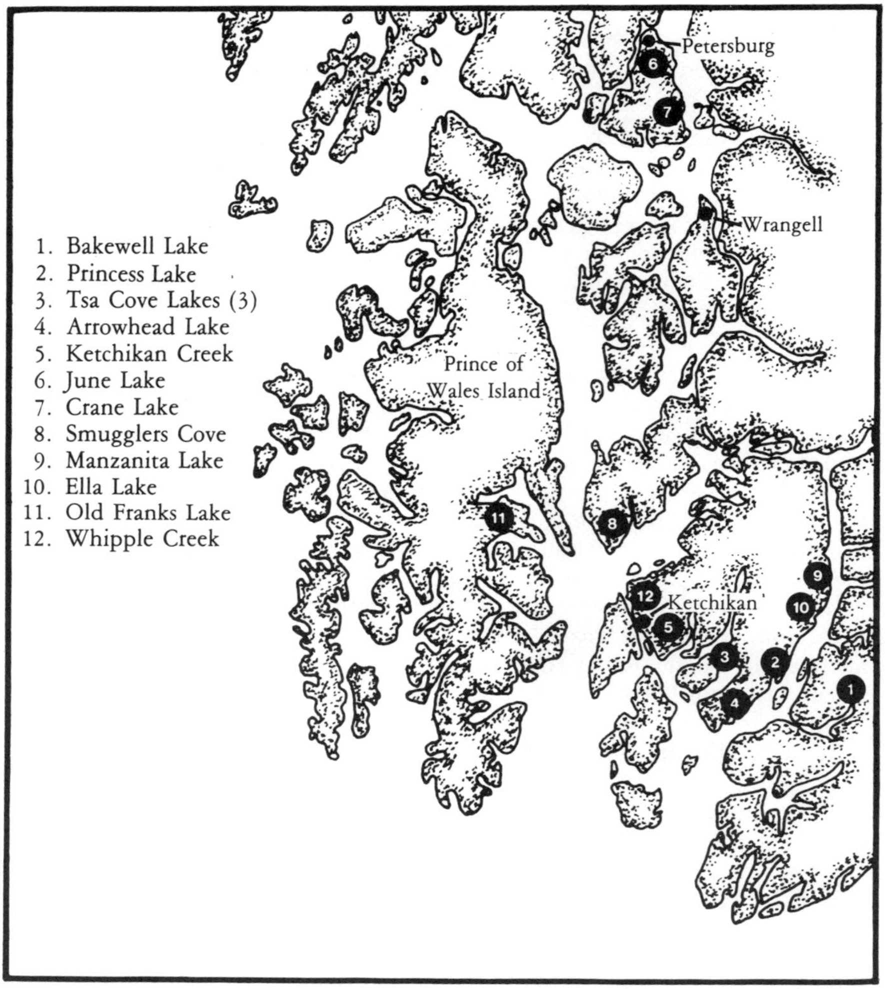

Map 14. Systems stocked with Salmon Fry from Deer Mountain Hatchery, 1955-1959. *Alaska Department of Fish and Game.*

Table 34. Releases of Salmon Fry from Deer Mountain Hatchery in Southern Southeast Alaska, 1955-1959

Release Date	Number of Fry	Number Per Pound	Egg Source	Disposition
COHO SALMON				
7/5/55	20,000	900	Soos Creek, Wn.	Whipple Creek
7/8/55	63,000	900	Soos Creek, Wn.	Ella Lake
7/12/55	63,000	900	Soos Creek, Wn.	Smugglers Cove Lake
7/21/55	19,500	500	Soos Creek, Wn.	Manzanita Lake
8/2/55	10,000	430	Soos Creek, Wn.	Whipple Creek
5/2/57	29,000	1,830	Ward Creek	Whipple Creek
5/22-5/23/57	90,000	1,830	Ward Creek	Smugglers Cove Lake
9/4/57	7,100	475	Ketchikan Creek	Smugglers Cove Lake
9/4/57	41,600	640	Ketchikan Creek	Smugglers Cove Lake
9/30-10/1/57	17,000	111	Ward Creek	Smugglers Cove Lake
10/2/57	3,330	111	Ketchikan Creek	Deer Creek
10/2/57	3,800	111	Ketchikan Creek	Schoenbar Creek[a]
10/21/57	250	95	Ketchikan Creek	Schoenbar Creek[b]
6/4-9/58	83,000	600-950	Ketchikan Creek and Ward Creek	Smugglers Cove Lake
8/30/58	11,000	160	Ward Creek	Unknown
9/2/58	3,500	114	Ward Creek	Unknown
9/15/58	2,500	78	Ward Creek	Unknown
5/25/59	45,000	1,500	Ketchikan Creek	TSA Lake #3
5/25/59	33,500	1,050	Ketchikan Creek	TSA Lake #1
8/4/59	16,200	550	Ketchikan Creek	Unknown
8/4/59	25,600	640	Ketchikan Creek	Whipple Creek
8/27/59	14,000	500	Ketchikan Creek	Smugglers Cove Lake
SOCKEYE SALMON				
5/4/59	19,600	1,635	Buschmann Creek	Arrowhead Lake[c]
5/25/59	103,300	2,650	Buschmann Creek	TSA Lake #1[d]
5/25/59	39,750	2,650	Buschmann Creek	TSA Lake #2
6/17/59	36,800	1,150	Buschmann Creek	June Lake[e]
7/3/59	42,400	1,766	Buschmann Creek	Crane Lake

(a) tributary to Ketchikan Creek
(b) all fin clipped.
(c) in East Behm Canal; lake treated with rotenone July 14, 1958.
(d) series of three lakes in George Inlet, treated with toxaphene and rotenone in 1958.
(e) both June and Crane lakes drain into Ideal Cove near Petersburg; both treated with rotenone.

Table 35. Releases of Salmon Fry from Deer Mountain Hatchery into Princess Lake, 1955-1959

Release Date	Species	Number of Fry	Number Per Pound	Egg Source
7/8/55	coho	4,500	900	Soos Creek, Wn.
4/15/58[a]	coho	30,450	1,450	Ward Creek
4/15/58	coho	14,500	1,600	Ward Creek
4/15/58	chinook	3,250	650	Soos Creek, Wn.
4/29/58	coho	45,000	1,550	Ward Creek
4/29/58	sockeye	90,000	2,900	Kah Shakes Lake
5/4/59	sockeye	45,000	2,900	Buschmann Creek
5/4/59	coho	45,000	1,500	Ketchikan Creek[b]

(a) Princess Lake treated with rotenone May, 1957.
(b) Ketchikan Creek brood stock was developed as indicated in text.

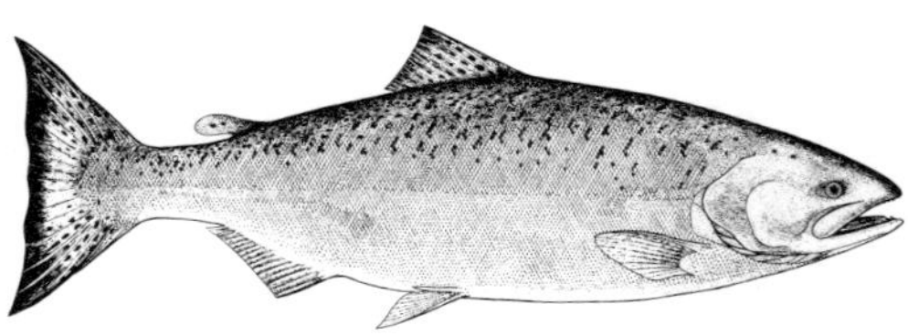

Notes

1. Letter from Director of U.S. Fish and Wildlife Service to Frederick Paul dated January 7, 1941, File "Sport Fishing in Alaska," General Records, Division of Alaska Fisheries, RG 22, NA.

2. F.F. Fish, "Progress Report, Artificial Propagation of Alaska Pink Salmon, Project IV, Sub-project 1," Western Fish Culture Investigations (National Marine Fisheries Service, Auke Bay Fisheries Laboratory: Auke Bay, Alaska).

3. Letter from Roger Burrows to Patricia Roppel dated October 1, 1979.

4. Letter from F.F. Fish to C.M. Mottley, Acting Chief, Section Eastern Inland Fishery Investigations dated November 26, 1947 (National Marine Fisheries Service, Auke Bay Fisheries Laboratory: Auke Bay, Alaska).

5. Will Brewington, "Summary of Pink Salmon Studies in Alaska," manuscript, 1956 (National Marine Fisheries Service, Auke Bay Fisheries Laboratory: Auke Bay, Alaska).

6. Ibid.

7. Alaska Department of Fisheries, *Annual Report for 1955,* Report #7 (Juneau, 1955), pp. 127-128.

8. For a comprehensive discussion of the Fraser Lake program after statehood see R.F. Blackett, "Establishment of Sockeye (Oncorhynchus nerka) and Chinook (O. tshawytscha) Salmon Runs at Fraser Lake, Kodiak Island, Alaska, *Journal of the Fisheries Research Board of Canada,* Vol. 36, No. 10, 1979, pp. 1265-1277.

9. Alaska Department of Fish and Game, *Annual Report for 1959,* Report No. 11 (Juneau: State of Alaska, 1959), pp. 32-33, 43.

10. Alaska Department of Fish and Game, *Annual Report for 1958,* Report No. 10 (Juneau: Territory of Alaska, 1958), p. 48.

11. Fry per square acre are given in the Alaska Department of Fish and Game, *Annual Report for 1959,* p. 28.

12. Number of fish, average length, average weight, percentage migrated, percentage holdover, and other data are given in the Alaska Department of Fish and Game, *Annual Report for 1959,* p. 28.

VIII. Epilogue

Alaska became a state on January 3, 1959. However, it did not assume control of its fisheries until January, 1960. The Alaska Statehood Act provided that the administration and management of the fish and wildlife resources of Alaska would be retained by the Federal Government under existing laws until the first day of the calendar year after the Secretary of Interior certified to Congress that the Alaska State Legislature had adequate provisions to take over these responsibilities.

Once control was assumed by the new state, Alaskans continued the philosophy of managing the existing runs of salmon. For the first ten years a small amount of salmon enhancement research continued at the Kitoi Bay Research Station, until it was destroyed in the 1964 earthquake; on Kodiak Island; and at the Deer Mountain Hatchery in Ketchikan.

To add to the knowledge of salmon enhancement, the National Marine Fisheries Service, through its Auke Bay Fisheries Laboratory, initiated projects in the 1960s to help define methods of fish culture which could be best adapted to the unique situations in Alaska. A facility, leased from the Territorial Sportsmen's Association, Juneau, and located on Auke Creek, has been used for various aspects of pink salmon research. The Little Port Walter field station, which stressed research on salmon biology after its inception in 1939, began in the late 1960s to work on coho salmon fish culture technology with emphasis on stocking barren lakes. Experiments on other species were conducted and in the late 1970s work was expanded to chinook salmon research. This work continues today.

The State of Alaska became actively involved in artificial production of salmon eleven years after gaining control of the fisheries. Alaskans again became concerned about declining salmon runs and urged the Alaska Legislature to create the Division of Fisheries Rehabilitation, Enhancement and Development (FRED) in 1971. This division of the Alaska Department of Fish and Game was charged with programs necessary to ensure perpetual increased production and use of food resources in Alaskan waters. The main thrust became a system of research and production salmon hatcheries in the state.

Six hatcheries have been built in Southeast Alaska including one near the old Klawak hatchery site and one in the remodeled building of the Deer Mountain hatchery. On Kodiak and Afognak islands a hatchery for pink and chum salmon now operates at Kitoi Bay near the original site. At long last on Karluk Lake, work is being done on sockeye enhancement using stream-side incubation boxes to eye eggs which are then planted in tributaries to the lake.

Salmon enhancement, or "aquaculture" as it is known today, has been expanded into parts of Alaska where hatcheries were unheard of in earlier times.

At present there are seven state hatcheries producing salmon in Southcentral Alaska, two of which are in Prince William Sound. Activities have been expanded into parts of the state which once would have been considered too cold for hatchery programs. The state has facilities near Dillingham, at Cold Bay, at Sikusuilaq Springs in the Kotzebue area, and near Fairbanks.

The biggest breakthrough in salmon fish culture since 1960 has been the development of a satisfactory fish food, one that is easy to obtain and ship, and meets the nutritional needs of small fish. It has been this which has been instrumental in reviving interest in salmon hatcheries. It has made it possible to rear fry to sizes which obtain much better survival rates. No longer is it necessary for hatchery personnel to locate sources of food and preserve it for future feedings. No longer do they worry if the fish will eat the food and if fish will thrive upon it.

However, unlike the hatcheries developed prior to statehood, few of the current facilities work with sockeye salmon. Outbreaks of a disease known as Infectious Hematopoietic Necrosis (IHN) have plagued hatchery operations. Best successes have been obtained from planting eyed eggs or planting fry soon after hatching into lake systems. The largest successes have been with pink salmon. Nearly 800,000 pink salmon returned as adults to the fisheries and to the state hatcheries in 1980. In addition 11,700 chinook, 22,400 coho, 17,300 chum and 22,400 sockeye were estimated to have returned that year. Production and estimated returns to state facilities continue to increase.

The State of Alaska ran the only salmon hatcheries from statehood until the mid-1970s. In 1974 legislation was passed to enable private non-profit organizations to build and operate hatcheries to provide salmon for the common property fisheries. But like the 1900 legislation, which brought hatcheries into the political arena, this 1974 legislation needed refining to be workable.

One significant addition to the legislation occurred when the 1976 Legislature created regional aquaculture associations. Since then five have been formed: Prince William Sound Aquaculture Association, Southern Southeast Regional Aquaculture Association, Northern Southeast Regional Aquaculture Association, Cook Inlet Aquaculture Association, and Lower Yukon/Kuskokwim Regional Aquaculture Association. A sixth will be the Aleutian/Pribilof Island Association. Regional associations comprise representatives of commercial fishermen and other user groups in the region including sport fishermen, processors, subsistence fishermen, and members of local communities.

These associations play an active role in planning and implementing salmon enhancement in their region through a Regional Planning Team made up of representatives of the Alaska Department of Fish and Game and the association. In addition many of the associations operate salmon hatcheries.

Twelve private non-profit hatchery permits had been issued by 1980 and 11 hatcheries were in operation of which five had returns of adult salmon. All are in Southeast Alaska except for two in Prince William Sound. The majority of these hatcheries are designed to produce pink and chum salmon. None work with sockeye salmon.

Regional aquaculture associations, in cooperation with the Alaska Department of Fish and Game and the U.S. Forest Service, have been involved in a program to enhance sockeye runs through lake fertilization to provide more food for fry in the system — something Henry O'Malley considered at Afognak Lake in the 1930s. Hugh Smith Lake, the site of the former Quadra hatchery, was fertilized in 1980 and 1981. Studies to determine the value of fertilizing Hetta Lake (site of the former Hetta hatchery) and Lake McDonald (site of the former Yes Bay hatchery) are in progress. Other sockeye systems, where no past hatchery operated, are being considered throughout the state. It is too early to determine if lake fertilization will be a viable alternative to hatcheries for sockeye salmon enhancement.

Political battles have continued during most legislative sessions since 1976 to refine the legislation permitting private non-profit aquaculture — a concept unique to Alaska. Regional associations and state agencies continue to be confronted with new, unforeseen problems brought about by this pioneer program. It remains to be seen whether this system of state and private non-profit hatcheries developed in the 1970s will be any more successful or have more longevity than the programs which were developed in the past.

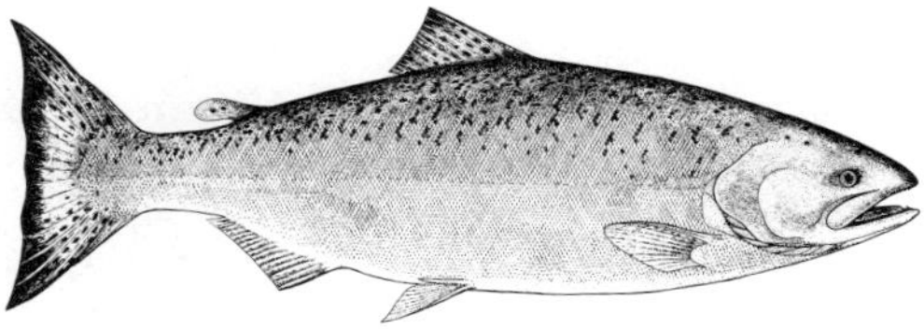

Selected Bibliography

An annotated bibliography was prepared simultaneously and will provide more detailed information than the selected bibliography.

Roppel, Patricia. *Alaska's Salmon and Trout Hatchery Programs — 1891-1959: An Annotated Bibliography*. Unpublished Manuscript, Auke Bay Fisheries Laboratory, Auke Bay, Alaska.

GOVERNMENT PUBLICATIONS

Alaska Governors. *Annual Reports,* 1897, 1899, 1901, 1909, 1911, 1918, 1921, 1924-1927. Washington: Government Printing Office, 1897-1927.

Alaska. *Laws of Alaska,* 1917, 1919, 1923. Juneau: Territory of Alaska, 1917, 1919, 1923.

Alaska Department of Fisheries. *Annual Reports,* 1950-1956. Juneau: Territory of Alaska, 1950-1956.

Alaska Department of Fish and Game. *Annual Reports,* 1957-1959. Juneau: Territory of Alaska, 1956-1959.

Alaska Territorial Fish Commission. Report of *Alaska Territorial Fish Commission for Years* 1920-1922. Juneau: Territory of Alaska, 1920-1923.

__________. *Special Report of Territorial Fish Commission to Legislature of Alaska.* Juneau: Alaska Daily Empire Print, 1923.

__________. *Report of the Alaska Territorial Fish Commission, Calendar Years 1923 and 1924.* Juneau: Alaska Daily Empire Print, 1924.

Bean, Tarleton H. "Report on the Salmon and Salmon Rivers of Alaska with Notes on the Conditions, Methods and Needs of the Salmon Fisheries," *Bulletin of the U.S. Fish Commission Vol. IX for 1889.* Washington: Government Printing Office, 1891.

Bower, Ward T. *Alaska Fisheries and Fur Industries* in 1918, 1919. Washington: Government Printing Office, 1919, 1920.

__________. *Alaska Fisheries and Fur-Seal Industries* in 1920-1936. Washington: Government Printing Office, 1921-1937.

Bower, Ward T. and Henry D. Aller. *Alaska Fisheries and Fur Industries* in 1914-1917. Washington: Government Printing Office, 1915-1918.

Bower, Ward T. and Harry Clifford Fassett. *Alaska Fisheries and Fur Industries in 1913.* Washington: Government Printing Office, 1914.

Brice, John J. *A Manual of Fish-Culture Based on the Methods of the United States Commission of Fish and Fisheries.* Washington: Government Printing Office, 1898.

Chamberlain, F.M. *Some Observations on Salmon and Trout in Alaska.* Washington: Government Printing Office, 1907.

Chamberlain, Fred M., and Ward T. Bower. "Fishery Industries," *Alaska Fisheries and Fur Industries in 1912.* Washington: Government Printing Office, 1913.

Cobb, John N. "The Commercial Fisheries of Alaska in 1906," *Report of the Commissioner of Fisheries for Fiscal Year 1905 and Special Papers.* Washington: Government Printing Office, 1906.

__________. "Report on the Fisheries of Alaska," *Report of the Commissioner of Fisheries, 1906 and Special Papers.* Washington: Government Printing Office, 1907.

__________. *The Salmon Fisheries of the Pacific Coast.* Washington: Government Printing Office, 1911.

__________. *Pacific Salmon Fisheries.* Washington: Government Printing Office, 1917, 1921, 1930.

Edfelt, Larry. *Statistical History of Alaska Salmon Catches.* Juneau: Alaska Department of Fish and Game, 1973.

Edginton, John and Robert Larson. *Revised Anadromous Stream Catalog of Southeast Alaska, Appendix A-District 1, Vol. III, Western Behm Canal and Ketchikan Area Survey Data.* Technical Data Report No. 30. Juneau: Alaska Department of Fish and Game, 1977.

Evermann, Barton Warren. *Alaska Fisheries and Fur Industries in 1911.* Washington: Government Printing Office, 1912.

Fassett, Harry Clifford. "The Karluk Hatchery," *Alaska Salmon Investigations in 1900 and 1901* by Jefferson Moser. Washington: Government Printing Office, 1902.

Gilbert, Charles H. "Age at Maturity of the Pacific Coast Salmon of the Genus Oncorhynchus," *Bulletin of the Bureau of Fisheries, Vol. XXXII for 1912.* Washington: Government Printing Office, 1914.

__________ and Willis H. Rich. "Investigations concerning Red Salmon Runs to the Karluk River, Alaska," *Bulletin of the U.S. Bureau of Fisheries Vol XLIII, 1927, Part II.* Washington: Government Printing Office, 1929.

Jones, Lester E. *Report of Alaska Investigations in 1914.* Washington: Government Printing Office, 1914.

Kutchin, Howard M. *Report on the Salmon Fisheries in Alaska* in 1897, 1898 1900-1905. Department of Commerce and Labor. Washington: Government Printing Office, 1898, 1899, 1901-1906.

Leach, Glen C. *Propagation and Distribution of Food Fishes, Fiscal Year* 1921-1933. Washington: Government Printing Office, 1922-1934.

__________ and M.C. James. *Propagation and Distribution of Food Fishes, Fiscal Year, 1934.* Washington: Government Printing Office, 1935.

Luttrell, J.K. "Report of Special Agent Luttrell," bound in *Salmon and Seal Fisheries, General Resources of Alaska, Vol. II, 1898.* Washington: Government Printing Office, 1893.

Marsh, Millard C. and John N. Cobb. *The Fisheries in Alaska* in 1907-1910. Washington: Government Printing Office, 1909-1911.

McDonald, Marshall. "Report on the Salmon Fisheries in Alaska," *Bulletin of the United States Fish Commission, Vol. XII for 1892.* Washington: Government Printing Office, 1894.

Moser, Jefferson F. *The Salmon and Salmon Fisheries of Alaska.* Washington: Government Printing Office, 1899.

__________. *Alaska Salmon Investigations in 1900 and 1901.* Washington: Government Printing Office, 1902.

O'Malley, Henry. *Artificial Propagation of the Salmons of the Pacific Coast.* Washington: Government Printing Office, 1920.

__________. "Report of the Commissioner for the Fiscal Year Ended June 30, 1923," *Report of the Commissioner of Fisheries for the Fiscal Year Ended June 30, 1923 and Appendices.* Washington: Government Printing Office, 1924.

Murray, Joseph. "Report of Joseph Murray, Special Treasury Agent, for the year 1895," bound in *Seal and Salmon Fisheries, General Resources of Alaska, Vol. II, 1898.* Washington: Government Printing Office, 1895.

__________. *Report on the Salmon Fisheries in Alaska-1894.* Washington: Government Printing Office, 1896.

Pracht, Max. *Report of the Agent for the Protection of Salmon Fisheries in Alaska.* Washington: Government Printing Office, 1893.

Rich, Willis H. and Edward M. Ball. "Statistical Review of the Alaska Salmon Fisheries, Part IV: Southeast Alaska," *Bulletin of the U.S. Bureau of Fisheries, Vol. XLVII.* Washington: Government Printing Office, 1935.

__________, and Arnie J. Suomela. "Salmon Tagging Experiments in Alaska, 1926," *Bulletin of the Bureau of Fisheries, Vol. XLIII, 1927, Part II.* Washington: Government Printing Office, 1927.

Stone, Livingston. "National Salmon Park," *Bulletin of the United States Fish Commission, Vol. XII for 1892.* Washington: Government Printing Office, 1894.

_________. "The Artificial Propagation on the Pacific Coast of the United States with Notes on the Natural History of the Quinnalt Salmon," *Bulletin of the U.S. Fish Commission, Vol. XVI for 1896.* Washington: Government Printing Office, 1897.

Sprague, A.J. *Report of the Territorial Fish Commissioner of Alaska, Territorial Legislature on H.C.R. 14, Session Laws, 1927-28.* Juneau: Territory of Alaska, 1928.

Tait, Howard D. *Kitoi Research Station, Its Development, Accomplishments and Future.* Alaska Department of Fish and Game Informational Leaflet No. 9. Juneau: State of Alaska, 1962.

Tingle, George R. *Report on the Salmon Fisheries of Alaska-1896.* Washington: Government Printing Office, 1897.

Titcomb, John W. "Fish Culture Practices in the United States Bureau of Fisheries," *Bulletin of the Bureau of Fisheries, Vol. XXVIII, 1908.* Washington: Government Printing Office, 1909.

United States Bureau of Fisheries. *Report of the U.S. Commissioner of Fisheries for Fiscal Year and Appendix.* 1912-1919, 1920-1923. Washington: Government Printing Office, 1914-1924.

U.S. Commission of Fish and Fisheries. *A Manual of Fish Culture, Based on the Methods of the United States Commission of Fish and Fisheries.* Revised Edition. Washington: Government Printing Office, 1900.

U.S. Congress, House, Committee on the Territories. *H.R. 9976, to Encourage Salmon Culture in Alaska,* Hearings, May 1, 1902, 57th Congress, 1st Session. Washington: Government Printing Office, 1902.

_________. *H.R. 13543 for the Protection and Regulation of the Fisheries in Alaska,* Hearings, March 7-9, 1906, 58th Congress, 1st Session. Washington: Government Printing Office, 1906.

_________. *Amendment of Laws Relating to Fisheries and Other Occupations in Alaska,* Hearings, April 19, 25, 28, May 2, 3, 5, 9, 25, and June 18, 1910. 60th Congress, 2nd Session. Washington: Government Printing Office, 1910.

_________. *Hearing on a Bill to Repeal the Law Allowing the Territory of Alaska to Tax Fisheries, H.R. 9627,* Hearings, March 28, 1916, 64th Congress, 1st Session. Washington: Government Printing Office, 1916.

U.S. Congress, House, Committee on the Merchant Marine and Fisheries. *H.R. 9528, A Bill for the Protection and Regulation and Conservation of the Fisheries of Alaska,* Hearings, 64th Congress, 1st Session. Washington: Government Printing Office, 1916.

U.S. Congress, House, Message from the President of the United States, *Transmitting a Communication from the Secretary of Commerce and Labor Submitting a Preliminary Report of the Alaska Salmon Commission.* 58th Congress, 2nd Session. Washington: Government Printing Office, 1904.

U.S. Congress, Senate. *Report of the Special Agent for the Protection of Alaska Salmon Fisheries.* 56th Congress, 1st Session. Washington: Government Printing Office, 1900.

U.S. Congress, Senate, Subcommittee of the Committee on Fisheries. *S. 5856, A Bill to Amend an Act for the Protection and Regulation of the Fisheries of Alaska.* Hearings, 62nd Congress, 2nd Session. Washington: Government Printing Office, 1912.

U.S. Department of Commerce and Labor. *Report of the Bureau of Fisheries, 1904.* Washington: Government Printing Office, 1905.

Wallich, Claudius. "A Method of Recording Egg Development for Use of Fish Culturists," *Report of the Commissioner of Fish and Fisheries for Year Ending June 30, 1900.* Washington: Government Printing Office, 1901.

Other Published Sources

American Fisheries Society. *Transactions of the American Fisheries Society,* (papers read at annual meetings; many papers on development of fish culture procedures).

Aro, K.V. *Transfer of Eggs and Young of Pacific Salmon within British Columbia,* Fisheries and Marine Service Technical Report No. 861. Nanaimo, British Columbia: Pacific Biological Station, 1979.

Bowen, J.T. "A History of Fish Culture as Related to the Development of Fishery Programs," *A Century of Fisheries in North America,* Norman Benson, editor. Washington: American Fisheries Society, 1979.

Bower, Ward T. "The Fisheries of Alaska," *Pacific Fisherman,* November, 1914.

Buschmann, C.H. "Hatcheries and the Disposition of the Salmon Packers." *Pacific Fisherman,* November, 1914.

Chamberlain, F.M. "Habits of Salmon." *Pacific Fisherman,* January, 1904.

Clarke, John J. "The Egg Layers." *Alaska Sportsman,* September, 1952.

Cobb, John N. "The Future of Alaska Fisheries." *Pacific Fisherman,* November, 1914.

Crutchfield, James E. and Givlio Pontecorvo. *The Pacific Fisheries: A Study of Irrational Conservation.* Baltimore: John Hopkins Press, 1969.

Evermann, Barton Warren, "Can the Alaska Salmon Fisheries be Saved." *Scientific Monthly,* February, 1921.

Gilbert, Charles H. *Summary of Contributions of the Life History of Sockeye Salmon,* Report of Commissioner of Fisheries, 1913, No. 1. Victoria, B.C.: 1914.

Jones, Robert D. "The Fortmann Salmon Hatcheries." *Alaska-Yukon Magazine,* November, 1907.

International Fishery Congress. *The United States Bureau of Fisheries: It's Establishment, Functions, Organization, Resources, Operations and Achievements.* Washington: Government Printing Office, undated.

O'Malley, Henry. "Best Methods of Spawning and Caring for Salmon Eggs." *Pacific Fisherman,* December, 1910.

Morton, William Markham. "Federal Fish Hatchery Memoirs," 1978-1979, Alaska Historical Library, Juneau.

Pratt, Daniel, "Farming the Seas." *Outing Magazine,* September, 1908.

Smith, E.A. "How the United States Government Can Save the Sockeye Salmon Industry." *Pacific Fisherman,* April, 1907.

Sprague, A.J. *The Alaska Salmon and Their Practical Propagation.* Juneau: Alaska Daily Empire Print, 1921.

Swift, Harry J. "The Klawak Hatchery." *Pacific Fisherman,* June, 1904.

Taylor, John P. "Eighteen Months at the Karluk Hatchery." *Alaska Magazine, January, 1964.*

Newspapers

The following newspapers were searched; for specific dates when articles appeared, check the annotated bibliography, Roppel, Alaska's Salmon and Trout Hatchery Programs — 1891-1959.

Alaska Daily Empire (Daily Alaska Empire after 1926) [Juneau]
Alaska Fisherman [Ketchikan]
Alaska Mining Record [Juneau]
Alaska Searchlight [Juneau]

Alaska Sentinel [Wrangell]
Alaska Weekly [Seattle]
The Alaskan [Sitka]
Cordova Daily Times
Daily Alaska Dispatch [Juneau]
Daily Miner [Ketchikan]
Daily Progressive Miner [Ketchikan]
Ketchikan, Alaska Chronicle
Ketchikan Miner
Ketchikan Daily News
Kodiak Mirror
The Mining Journal [Ketchikan]
Seward Daily Gateway
Wrangell Sentinel

Unpublished Sources

The major portion of the original material used is located at the National Archives in Washington D.C. These are the records of the United States Fish and Wildlife Service, Record Group 22. Within these records the Records of the Division of Alaska Fisheries and the Division of Fish Culture were primarily consulted.

Other original sources include logbooks from several of the hatcheries: the Yes Bay logbooks, 1906-1934 are at the Alaska Historical Library, Juneau; the Hetta logbooks, 1910-1918, and the Quadra logbooks, 1908-1935, were donated to the author by Mark Meyer and subsequently donated to the Alaska Historical Library.

The Records of the Office of the Territorial Governors in the Alaska State Archives in Juneau contain the best sources on activities of the Alaska Territorial Fish Commission.

The Auke Bay Fisheries Laboratory (National Marine Fisheries Service) records stored at the Federal Record Center, Seattle, contain information on the Karluk hatchery.

The diaries of John C. Callbreath are located at the University of Washington library and the papers of John N. Cobb are in that university's archives.

Information was provided through interviews with John Olafson, Alphonse Kemmerich and Mark Meyer as well as through correspondence with Kemmerich, Meyer, Harry D. Baer, Vincent Boucher, Clare Ritter, William Paul Sr., Helen Wentworth, and Emery Tobin.

Patricia Roppel

A graduate of Oregon State University, Patricia Roppel came to Alaska as a teacher. She taught high school for six years after her arrival in 1959. Since then Mrs. Roppel has developed a keen interest in Alaskan history and her articles have appeared frequently in Alaskan magazines. She was awarded the Alaska Historical Society's "Beaver Log Award" in 1974; and was named "Alaska Historian of the Year" in 1977 for her continued excellence in historical research. She serves as a commissioner on the Alaska Historical Commission.

Mrs. Roppel's interest in the historical aspects of hatchery operations has been augmented by her involvement in current aquaculture programs. She has written articles and prepared reports for the Division of Fisheries Rehabilitation, Enhancement and Development of the Alaska Department of Fish and Game. As corporate secretary of the Southern Southeast Regional Aquaculture Association, she has been actively involved since its inception in 1976. The non-profit aquaculture association currently operates a hatchery near her hometown of Ketchikan.

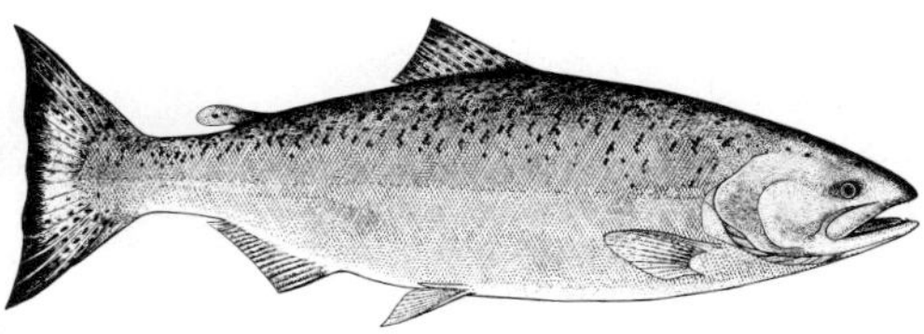

Index